# STRATEGIC INFORMATION WARFARE RISING

**Roger C. Molander/Peter A. Wilson/David A. Mussington/Richard F. Mesic**

Prepared for the
Office of the Secretary of Defense

National Defense Research Institute

RAND

This report summarizes research performed by RAND for the Office of the Assistant Secretary of Defense (Command, Control, Communications and Intelligence) in response to a request from the Office of the Deputy Secretary of Defense. The objective of this effort was to derive a framework for policy and strategy decisionmaking on problems raised by the emerging potential of Strategic Information Warfare.

This study was undertaken in recognition that future U.S. national security strategy is likely to be profoundly affected by the ongoing rapid evolution of cyberspace—the Global Information Infrastructure (GII)—and, thus by the growing dependence of the U.S. military and other national institutions and infrastructures on potentially vulnerable elements of the U.S. national information infrastructure.

This report should be of special interest to those who are exploring the effect of the information revolution on strategic warfare, and to those who are concerned with ensuring the security of information-dependent infrastructures. It should also be of interest to those segments of the U.S. and the international security community that are concerned with the post–Cold War evolution of military and national security strategy, especially strategy changes driven wholly or in part by the evolution of, and possible revolutions in, information technology.

The research reported here was accomplished within the Acquisition and Technology Policy Center of RAND's National Defense Research Institute, a federally funded research and development center sponsored by the Office of the Secretary of Defense, the Joint Chiefs of Staff, Unified Commands, and the defense agencies. It builds on an earlier and ongoing body of research within the center on the national security implications of the information revolution.

# CONTENTS

Preface . . . . . . . . . . . . . . . . . . . . . . . . . . . . . . . . . . . . . . . . . . . . . . . . . . . iii

Figures . . . . . . . . . . . . . . . . . . . . . . . . . . . . . . . . . . . . . . . . . . . . . . . . . . vii

Tables . . . . . . . . . . . . . . . . . . . . . . . . . . . . . . . . . . . . . . . . . . . . . . . . . . . ix

Summary . . . . . . . . . . . . . . . . . . . . . . . . . . . . . . . . . . . . . . . . . . . . . . . . . . xi

Chapter One
WHAT IS STRATEGIC INFORMATION WARFARE? . . . . . . . . . . . . . . . . . . 1
Introduction . . . . . . . . . . . . . . . . . . . . . . . . . . . . . . . . . . . . . . . . . . . . 1
What Is Strategic Warfare? . . . . . . . . . . . . . . . . . . . . . . . . . . . . . . . . . . 2
What Is Information Warfare? . . . . . . . . . . . . . . . . . . . . . . . . . . . . . . . . 3
The History and Future of Strategic Information Warfare . . . . . . . . . . . . 5

Chapter Two
THE STRATEGIC INFORMATION WARFARE . . . . . . . . . . . . . . . . . . . . . . 9
The Need for New Decisionmaking Frameworks . . . . . . . . . . . . . . . . . . . 9
An Evolving Series of Frameworks . . . . . . . . . . . . . . . . . . . . . . . . . . . . . 10
Initial Formulation of a First-Generation SIW Strategy And Policy
    Decisionmaking Framework . . . . . . . . . . . . . . . . . . . . . . . . . . . . . . 11

Chapter Three
KEY DIMENSIONS OF THE SIW ENVIRONMENT . . . . . . . . . . . . . . . . . . . 15
From Defining Feature to Key Dimensions . . . . . . . . . . . . . . . . . . . . . . . 15

Chapter Four
KEY STRATEGY AND POLICY ISSUES . . . . . . . . . . . . . . . . . . . . . . . . . . 17
The Issue Menu . . . . . . . . . . . . . . . . . . . . . . . . . . . . . . . . . . . . . . . . . . 17
Locus of Responsibility and Authority . . . . . . . . . . . . . . . . . . . . . . . . . . 19
Tactical Warning, Attack Assessment, and Emergency Response . . . . . . . . 19
Vulnerability Assessments . . . . . . . . . . . . . . . . . . . . . . . . . . . . . . . . . . . 20
Declaratory Policy . . . . . . . . . . . . . . . . . . . . . . . . . . . . . . . . . . . . . . . . 21
International Information Sharing and Cooperation . . . . . . . . . . . . . . . . . 21
Investment Strategy . . . . . . . . . . . . . . . . . . . . . . . . . . . . . . . . . . . . . . . 22

Chapter Five
CURRENT STATE OF FIRST-GENERATION SIW . . . . . . . . . . . . . . . . . . . 25
Assessing An Embryonic Concept . . . . . . . . . . . . . . . . . . . . . . . . . . . . . 25
Key Factors in SIW Development to Date . . . . . . . . . . . . . . . . . . . . . . . . 26

Assessing Current Levels of Offensive SIW Capability . . . . . . . . . . . . . . .    27
Assessing Current Levels of Defensive SIW Capability . . . . . . . . . . . . . . .    28
A Preliminary Assessment of Where We Are . . . . . . . . . . . . . . . . . . . . . . .    30

Chapter Six
ALTERNATIVE FIRST-GENERATION SIW END STATES . . . . . . . . . . . . . .    33
Introduction . . . . . . . . . . . . . . . . . . . . . . . . . . . . . . . . . . . . . . . . . . . . . . .    33
An Initial Array of Possible End States . . . . . . . . . . . . . . . . . . . . . . . . . . .    33
The First in an Evolving Series of Frameworks . . . . . . . . . . . . . . . . . . . . .    36

Chapter Seven
AN EVOLVING SERIES OF FRAMEWORKS . . . . . . . . . . . . . . . . . . . . . . . .    39
Introduction . . . . . . . . . . . . . . . . . . . . . . . . . . . . . . . . . . . . . . . . . . . . . . . .    39
Major Potential Perturbations . . . . . . . . . . . . . . . . . . . . . . . . . . . . . . . . . .    39
The Framework as a Means of Shaping the Future . . . . . . . . . . . . . . . . . .    42

Chapter Eight
ALTERNATIVE ACTION PLANS . . . . . . . . . . . . . . . . . . . . . . . . . . . . . . . . .    43

Appendix A
EXEMPLARY FIRST- AND SECOND-GENERATION SIW
ESCALATION SCENARIOS . . . . . . . . . . . . . . . . . . . . . . . . . . . . . . . . . . . .    47

Appendix B
HOW TO USE THIS TOOL . . . . . . . . . . . . . . . . . . . . . . . . . . . . . . . . . . . . .    59

Appendix C
EXEMPLARY SIW SCENARIOS . . . . . . . . . . . . . . . . . . . . . . . . . . . . . . . . .    75

Appendix D
THE STRATEGIC NUCLEAR WARFARE FRAMEWORK PROBLEM . . . . . . .    77

S.1. Asymmetric Strategies That Might Be Sought by Future U.S. Regional Adversaries .................................... xii
S.2. Two Concepts of Strategic Information Warfare ................ xiii
S.3. Steps in Designing a First-Generation SIW Strategy and Policy Decisionmaking Framework ............................ xv
1.1. Strategic Information Warfare ............................ 1
1.2. Asymmetric Strategies That Might Be Sought by Future U.S. Regional Adversaries .................................... 4
1.3. Symmetric Strategies That Might Be Presented by Future U.S. Peer Competitors .................................... 4
1.4. Two Concepts of Strategic Information Warfare ................ 6
2.1. Designing a First-Generation SIW Strategy and Policy Decisionmaking Framework ............................ 12
6.1. Where Are We (X) and Where Should We Be Going (A-D)? ......... 36
6.2. Where Might We Be in the Future? ......................... 37
A.1. Overview SIW Escalation Chart ........................... 48
A.2. China-Taiwan 2010 Scenario Escalation Chart ................ 52
A.3. Russia 2000–2010 SIW Scenario Escalation Chart ............... 55

S.1. Defining Features, Consequences, and Key Dimensions of the
SIW Environment . . . . . . . . . . . . . . . . . . . . . . . . . . . . . . . . . . . . . . . . xvi
S.2. Alternative Action Plans . . . . . . . . . . . . . . . . . . . . . . . . . . . . . . . . . . xxv
3.1. Defining Features, Consequences, and Key Dimensions of the
SIW Environment . . . . . . . . . . . . . . . . . . . . . . . . . . . . . . . . . . . . . . . . 16
6.1. Alternative Strategic Information Warfare (SIW) Asymptotic End
States . . . . . . . . . . . . . . . . . . . . . . . . . . . . . . . . . . . . . . . . . . . . . . . . . 34
8.1. Alternative Action Plans . . . . . . . . . . . . . . . . . . . . . . . . . . . . . . . . . . 44
B.1. Models of Tactical Warning and Alert Structure . . . . . . . . . . . . . . . . 64
B.2. Models of Infrastructure Protection Research and Development
Resource . . . . . . . . . . . . . . . . . . . . . . . . . . . . . . . . . . . . . . . . . . . . . . . 68
D.1. Alternative Strategic Nuclear Warfare Asymptotic End States . . . . . . . 78
D.2. Alternative Strategic Nuclear Warfare Asymptotic End States:
Inaugural Period . . . . . . . . . . . . . . . . . . . . . . . . . . . . . . . . . . . . . . . . 79
D.3. Alternative Strategic Nuclear Warfare Asymptotic End States:
1949–1954 . . . . . . . . . . . . . . . . . . . . . . . . . . . . . . . . . . . . . . . . . . . . . 80
D.4. Alternative Strategic Nuclear Warfare Asymptotic End States:
1962–1964 . . . . . . . . . . . . . . . . . . . . . . . . . . . . . . . . . . . . . . . . . . . . . 81
D.5. Alternative Strategic Nuclear Warfare Asymptotic End States:
Mature Outline 1997 . . . . . . . . . . . . . . . . . . . . . . . . . . . . . . . . . . . . . 82

## WHAT IS STRATEGIC INFORMATION WARFARE?

In the future, the possibility exists that adversaries might exploit the tools and techniques of the Information Revolution to hold at risk (not for destruction, but for large-scale or massive disruption) key national strategic assets such as elements of various key national infrastructure sectors, such as energy, telecommunications, transportation, and finance). This potential danger constitutes the principal aspect of the Strategic Information Warfare (SIW) environment addressed in this report.

Both regional adversaries and peer competitors may find SIW tools and techniques useful in challenging the United States, its allies, and/or its interests. SIW weapons may find their highest utility in the near-term in "*asymmetric" strategies* employed by regional adversaries (see Figure S.1). Such adversaries might seek to avoid directly challenging U.S. conventional battlefield superiority through a more indirect attack (or threat) involving some combination of nuclear, chemical, biological, highly advanced conventional, and SIW instruments.

SIW tools and techniques present a two-pronged threat to U.S. security:

1. **a threat to U.S. national economic security.** Key national infrastructure targets could be at risk to such massive disruption that a successful attack on one or more infrastructures could produce a strategically significant result, including public loss of confidence in the delivery of services from those infrastructures.

2. **a threat against the U.S. national military strategy.** The possibility exists that a regional adversary might use SIW threats or attacks to deter or disrupt U.S. power projection plans in a regional crisis. Targets of concern include infrastructures in the United States vital to overseas force deployment, and comparable targets in allied countries. A key ally or coalition member under such an attack might refuse to join a coalition—or worse, quit a coalition in the middle of a war.

In the history of strategic warfare, it is hard to find a conflict worthy of the label "strategic" that did not manifest some important information component. Sun Tzu, for example, recommended the creative use of information to achieve strategic objectives while avoiding conflict. It is also noteworthy that one could undoubtedly produce a list of historical instances in which fundamental changes in technology produced fundamental changes in the information component of strategic warfare.

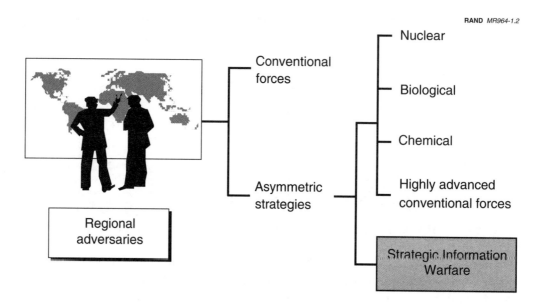

**Figure S.1—Asymmetric Strategies That Might Be Sought by Future U.S. Regional Adversaries**

Yet the potential impact of the Information Revolution on strategic warfare may be unprecedented. In the past, SIW may have played largely a subordinate role in strategic warfare—in early times in the strategic impact of conventional armies and navies, and later through airplanes, rockets, or nuclear weapons. However, SIW might play a much greater role in such warfare in the wake of the Information Revolution. Furthermore, the potential impact of the Information Revolution on the vulnerability of key national infrastructures and other strategic assets may over time give rise to a new kind of information-centered strategic warfare based on completely different time lines, and worth consideration independent of other potential facets of strategic warfare such as those portrayed in Figure S.1.

SIW thus might be conceptualized in the following terms (see Figure S.2):

1. **First-Generation SIW.** SIW as one of several components of future strategic warfare, broadly conceptualized as being orchestrated through a number of strategic warfare instruments (as indicated in Figure S.1).

2. **Second-Generation SIW.** SIW as a free-standing, fundamentally new type of strategic warfare spawned by the Information Revolution, possibly implemented in newly prominent strategic warfare arenas (for example, economic) and on time lines (for example, years versus days, weeks, or months) than those generally, or at least recently, ascribed to strategic warfare.

For established powers such as the United States, the authors tend to believe that first-generation SIW is the more likely form of strategic warfare to be initially mani-

RAND *MR963-4*

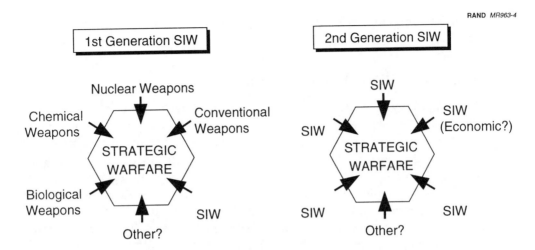

Figure S.2—Two Concepts of Strategic Information Warfare

fested. This proposition, is however, arguable. The United States, for example, might find itself in a situation in the near future in which it chooses to exploit its current information technology (IT) advantages and employ second-generation SIW to prevail in a crisis that otherwise would have led to troop deployments and almost certain high numbers of casualties.

For less-developed nations, which may not possess any other effective strategic warfare instruments, second-generation SIW may be more immediately attractive. In fact, second-generation SIW use by or against lesser powers might follow close on the heels of the demonstration of first-generation SIW.

## THE NEED FOR NEW DECISIONMAKING FRAMEWORKS

According to the project description for this study, "The goal of this project is to formulate a common DoD strategy and policy framework for addressing the challenge of strategic information warfare." But what is a strategy and policy decisionmaking framework? A decisionmaking framework is likely to be a series of relatively simple steps, or a process, that presents the strategy, policy, and related issues that need to be addressed in some particular arena in a logical architecture, and along a logical path in a manner that facilitates decisionmaking on those issues.

New strategy and policy decisionmaking frameworks are born in the crucible of necessity (or perceived possible imminent necessity) manifested when a specific problem area (1) appears to demand action (or might soon demand action) and (2) is of such a nature that no readily applicable decisionmaking framework to forge an implementable action plan is available.

In some situations, an older decisionmaking framework may have been tested for its applicability to the needs of the subject problem area and been found wanting. Those who favor formulating the subject area as a rapidly evolving old problem area versus a new problem area may, in fact, have championed use of such an older framework. Failed attempts to apply an older decisionmaking framework may even

xiv    Strategic Information Warfare Rising

have contributed to a delay in the more forthright expression of the need for a new framework.

## AN EVOLVING SERIES OF FRAMEWORKS

The history of the carrying out of the above-mentioned tasks can be characterized as an initial search for a single, temporally stable framework to serve the stated function for SIW that soon concluded that the concept of a *single framework* at this stage of development was illusory. Rather, the correct construct for responding to a new strategic warfare component—one truly worthy of the label "strategic" and opposed to just another "strategic warfare wannabe"—would have to be dynamic, and capable of responding to ongoing changes in both the international security and IT environments. The correct construct would have to be (1) *an evolving series of frameworks*, recognizing and accepting the evolution-like "punctuated equilibrium" realities of convening and executing strategy and policy decisionmaking processes, and (2) *a process* that recognizes and supports the dynamic and highly evolutionary character of such a construct (especially in its early stages).

## AN INITIAL FORMULATION

A primary objective in this conceptualization of the SIW decisionmaking framework problem is that the initial formulation of the framework be one that can evolve in response to changes in its environment. It needs to have evolutionary potential, rather than being a temporary expedient that got decisionmaking going, but did not have much utility thereafter.

Because there is no precursor framework in the SIW area, the initial version of the framework will attract attention from stakeholders interested in the future of the Information Revolution and from the media. The process of designing an associated inaugural first-generation SIW decisionmaking framework—a process that *constitutes* the framework—can therefore be divided into the following distinct steps (see Figure S.3):

1. **Key dimensions of the SIW environment.** Gain an understanding of the key dimensions of the future first-generation SIW "environment" or "battlespace," that is, those dimensions of that environment that might, in principle, be influenced (presumably in some favorable direction) by effective near-term strategy and policy decisionmaking. Achieve this objective through (1) the identification of the principal defining features of first-generation SIW within a spectrum of plausible first-generation SIW contexts and (2) the selection of those features that might be cast as key dimensions amenable to change as described above.

2. **Key strategy and policy issues.** Identify those key strategy and policy issues (and such other issues as organizational issues) germane to the first-generation SIW problem (that is, issues for which near-term decisionmaking could influence the key dimensions of the SIW environment identified above).

RAND *MR964-S.3*

```
┌─────────┐                              ┌─────────┐
│ Step 1  │                              │ Step 2  │
└─────────┘                              └─────────┘
  ┌──────────────────────────────┐  ┌──────────────────────────────┐
  │  Defining features  ⇨  Key   │⇨ │    Key strategy and          │
  │                    dimensions │  │    policy issues             │
  └──────────────────────────────┘  └──────────────────────────────┘

        ┌─────────┐                        ┌─────────┐
        │ Step 3  │                        │ Step 4  │
        └─────────┘                        └─────────┘
    ┌──────────────────────────┐    ┌──────────────────────────────┐
 ⇨  │  Current state of        │ ⇨  │  Alternative first-generation │
    │  first-generation SIW    │    │  SIW end states              │
    └──────────────────────────┘    └──────────────────────────────┘

              ┌─────────┐
              │ Step 5  │
              └─────────┘
        ┌──────────────────────────┐
     ⇨  │  Alternative action plans │
        └──────────────────────────┘
```

Figure S.3—Steps in Designing a First-Generation SIW Strategy and Policy
Decisionmaking Framework

3. **Current state of First-Generation SIW.** Assess the current state of first-generation SIW in terms of absolute and relative offensive and defensive SIW capabilities.

4. **Alternative First-Generation SIW "end states."** In light of the above-mentioned first-generation SIW contexts and scenarios, craft a set of (plausible and potentially desirable) alternative first-generation SIW "end states"—expressed in terms of the above mentioned key dimensions of the first-generation SIW environment.

5. **Alternative action plans.** Array the key SIW strategy and policy issues against each of the alternative end states, and conceptualize action plans for moving toward one or more of these end states.

Any such framework will have to be continually tested and evaluated against emerging contingencies. It should be recognized, however, that it may be hard to achieve a sustained high level of comfort concerning the viability of any framework until the related IT and international security environments are less dynamic. Further details on the five steps shown in Figure S.3 are provided below.

## KEY DIMENSIONS OF THE SIW ENVIRONMENT

As previously noted, the key dimensions of the SIW environment are obtained by identifying the defining features of the SIW environment, and asking which of these can be potentially influenced in some favorable direction by well-conceived strategy and policy decisionmaking. These dimensions (see Table S.1) thus constitute the basic factors in the SIW setting that influence attainable objectives relating to SIW, and the relationships between purposeful action by nations (and other "actors") and changes in the shape of the SIW environment itself.

## KEY STRATEGY AND POLICY ISSUES

SIW presents a broad and complex spectrum of issues and challenges to existing decisionmaking processes. Thus, it is clear that some sequencing in taking up these issues nationally and internationally is appropriate. The key strategy and policy issues identified in this study can therefore be roughly characterized in terms of three categories:

"Low-Hanging Fruit." Those issues that could be moved to closure nationally (and, in some cases, internationally) without undue difficulty once suitable processes are identified or established. Issues in this category (with sample alternatives) are

- **Locus of responsibility and authority.** Who should have the lead responsibility—government (and, if so, who within the government) and/or industry (and, if so, who within the key infrastructures in the U.S. national response to the SIW threat?

  — Federal government leadership with a national security focus.

  — Federal government leadership with a law-enforcement focus (for example, Department of Justice leadership)

  — Joint international government leadership with a national security focus

  — Joint international government leadership with an law-enforcement focus

  — International industry leadership with government support.

### Table S.1

#### Defining Features, Consequences, and Key Dimensions of the SIW Environment

| Defining Features | Consequences | Key Dimensions |
|---|---|---|
| Entry cost low | May be many actors in the SIW battlespace | Number of offensive SIW players |
| Strategic intelligence on threat unavailable | Identity and capabilities of potential adversaries may be unclear | Number of offensive SIW players |
| Tactical warning difficult | May not know attack is under way | Tactical warning capability |
| Attack assessment difficult | May not know perpetrator or targets | Attack assessment capability, including perpetrator identity |
| Damage assessment difficult | May not know full implications of the attack | Damage assessment capability |
| Traditional boundaries blurred | May not know who has various responsibilities before, during, or after attack | N/A |
| Weapon effects uncertain | Both attacker and defender may be uncertain about weapon effects | Uncertainty in weapon effects |
| Infrastructure vulnerabilities uncertain but suspect | U.S. homeland may not be a sanctuary; vulnerable partners could make sustaining coalitions more difficult | Degree of SIW vulnerability |

- **Tactical warning, attack assessment, and emergency response.** How should the United States (and the world), including its governments and its industry, organize to develop and implement capabilities and procedures to sense and respond to SIW threats?

  — A government-led national security–oriented model (called a National Infrastructure Condition [NICON] model)

  — A government-led law-enforcement-oriented model (called a counterterrorism model)

  — A Centers for Disease Control and Prevention (CDC) model

  — An industry-led model.

- **Vulnerability assessments.** By what means and mechanisms of government and industry cooperation should a vulnerability assessment of key U.S. national infrastructures be undertaken?

  — A government-led (these could include for example, DoD-led) assessment of U.S. vulnerabilities

  — A joint public and private sector effort involving the United States and other key nations (for example, G-7[1] and/or potential SIW peer competitors)

  — An international public-private partnership, such as the CDC and the World Health Organization (WHO)

  — An industry-led and government-assisted assessment.

- **Declaratory policy on SIW use.** What should U.S. government declaratory policy be on the use of SIW and the relationship between the use of SIW and other strategic military and economic instruments?

  — Retaliation principally in kind for any SIW attack

  — Retaliation principally by non-SIW military means in response to such an attack

  — Retaliation by economic means, possibly including economically oriented SIW means, in response to such an attack

  — Complete ambiguity as to how the United States would respond to such an attack.

## Tough Issues to Be Faced Now

Urgent but contentious issues related to the inaugural charting of long-term SIW-related national goals and strategy. Examples of these issues (with alternatives) include

- **Research and development (R&D) investment strategy.** What investment strategy should the United States pursue for (1) monitoring, perpetrator identifica-

---

[1]The G-7 is the name applied to the seven largest industrial democracies (United States, Canada, France, Germany, Great Britain, Italy, and Japan) which meet annually at the level of chiefs of state.

tion, and perpetrator "trackback" techniques, (2) attack assessment techniques, (3) defense and reconstitution techniques, and (4) damage assessment techniques?

— A government-led national security–oriented model (called a National Infrastructure Condition (NICON) model

— No significant international SIW cooperation

— Limited international cooperation focused on defensive techniques (such as the G-7 model)

— Broad international cooperation organized through existing multinational security arrangements (for example, the NATO model)

— Broad international cooperation organized through global arrangements(such as the WHO model).

— Broad voluntary international cooperation.

• **International information sharing and cooperation.** What principles should guide international collaboration (in particular with allies and coalition partners) in the SIW domain? Is there an SIW parallel to extended deterrence?  To extended defense?

— National security–oriented network protection goals

— Coordinated defensive R&D with allies

— International proscriptions on offensive SIW R&D

— Private sector or market-driven focus.

## Deferred Issues

Issues that are not yet ready, for example, because of technical uncertainties to be taken to closure, or, worse, issues that are taken to closure prematurely, possibly producing "bad" strategy or policy decisions that would be hard to undo.  Issues in this category include

• **Intragovernmental and intergovernmental cooperation on politically sensitive privacy issues.**  This subject needs to be included in any discussion of SIW, but more detail is needed on how privacy rights would be protected under specific strategies and policies.

• **Minimum essential information infrastructure (MEII).**  More analytical and conceptual work is needed to determine whether the MEII concept (a system providing a minimal level of communications access and services to critical governmental and societal user communities) is at all feasible from both a technical and cost standpoint.

• **Encryption policy.**  SIW is just one of the many issue areas that need to be "brought to the table" when the United States and the international community chart long-term encryption-related goals and strategies.

Each of these areas requires sensitive treatment.  In turn, each of them overlaps with other elements of a comprehensive approach to addressing SIW policy concerns. The notion that an action plan for addressing SIW vulnerabilities requires that tradeoffs be made among different factors is central to the unprecedented uncertainties of the cyberspace environment.  The next section addresses defensive and offensive SIW issues that are significant to SIW action plans and policy implementation.

## CURRENT STATE OF FIRST-GENERATION SIW

A macro assessment of the current state of first-generation SIW in terms of absolute and relative offensive and defensive SIW capabilities of the United States and other nations (or other parties) would be difficult to do, even at a classified level.  The current dynamic character of the Information Revolution and the embryonic character of SIW as a potential political-military instrument both argue for caution in making such an assessment, classified or unclassified, at present and for the foreseeable future.

Principal SIW assessment issues from the U.S. perspective are

- The extent to which hostile SIW powers already exist and the degree to which they can seriously harm the United States with SIW attacks

- The extent of current U.S. offensive SIW capability compared with that of other nations (whether foe, neutral, or friend)—whether overt or covert—in preventive, preemptive, or retaliatory SIW actions.

To address this issue, the difficult task of evaluating offensive and defensive SIW capabilities must be broached.

The United States, as the global leader in the development and exploitation of information systems, has the most potential to be an offensive SIW "superpower."  Any lesser assessment of U.S. SIW potential compared with the SIW potential of other nations would be judged as laughable by those nations that are just beginning to speculate about the significance that SIW instruments may have in future conflicts. But how far has this U.S. SIW potential been exploited?  How fast could it be exploited if the United States were to make a strong national commitment to the urgent development of offensive SIW capabilities?

On the offensive side, the current U.S. experience with information operations is as a supporting but relatively low profile element of U.S. military strategy and doctrine. The U.S. has well-developed and successful offensive command and control warfare ($C^2W$), electronic warfare (EW), and other information warfare (IW) capabilities (for example, SOCOM is a master of psychological operations (psyops), and the military services develop and operate electronic warfare systems, as manifested in the large-scale use of $C^2W$ and the suppression of enemy air defenses in the Persian Gulf War), but these can hardly be characterized as "strategic" in the sense of this report. Offensive first-generation SIW, which by definition has the potential to hold at risk a country's "central nervous system" (its critical infrastructure networks), is a much more sensitive undertaking than are "information operations" as supporting missions in conventional warfare. It is one thing to target military leadership, communications, and radar; it is quite another to target public utilities that, among other purposes, provide power to hospitals.

The sensitivities of our friends and allies and the political-military capital that might accrue to possible adversaries from an increasingly open emphasis on U.S. offensive SIW initiatives have largely kept more definitive information on these capabilities from being revealed. Although some U.S. SIW offensive capability exists, its full potential is politically and militarily sensitive.

Beyond being a leading contender in augmenting its existing arsenal with offensive SIW capabilities, the United States, by virtue of its political, economic, and technological position in the world, is also a natural target for SIW attack. The United States leads the world in the development and application of information technologies and has a complex society and economy that are very dependent on information systems. It is geographically protected and currently has the world's most formidable conventional military capabilities. If the United States were to be defeated or thwarted militarily in the near future, it will probably be because of the successful use of an asymmetric strategy by an enemy seeking to avoid a direct military confrontation.

The first logical step in understanding SIW defensive implications is to conduct a review of potential U.S. vulnerabilities to conceivable SIW attacks across a broad spectrum of threats and scenarios. Unfortunately (or fortunately), we have very little "real-world" experience on which to base such an assessment. There have been a number of natural events (such as storms and earthquakes), human errors (in software and control), and other purposeful mischief (such as hobbyist hackers and criminals) that suggest that things can go wrong in various national infrastructures, occasionally on an impressive scale. But none of these past events has been "strategic" in its impact, nor do any appear to have been strategic in intent.

One obvious problem with this paucity of defensive SIW-related experience is in relating cause and effect: Have we escaped SIW attacks because certain undetected attempts were not successful or because no attempt has been made yet?

Although a great deal of uncertainty surrounds the future vulnerability of information infrastructures, a number of trends can be observed that seem to point toward an expanded dependence on less secure networking concepts. In particular, the widespread adoption of open network standards and technologies means that the industries and applications delivered via cyberspace may become more vulnerable to single-point failures. The growth of electronic commerce, the prospective expansion of electronic stored value payment systems (called cyberpayment), and plans for the delivery of critical services (such as telemedicine and government communications) over the GII all present potential targets for an SIW attack.

The defensive SIW assessment thus involves an assessment of information infrastructure vulnerability, threat potential, and vulnerability consequences. However, these assessments also have their problems. Existing information infrastructure systems are complex, dynamic, flexible, and interdependent. They are also public and private, and military and commercial. Some (such as those used in banking) have been "hardened" by design because of the potential risk and cost of compromise. Others have evolved in a more benign environment with functions not related to threat (for example, cost, accessibility, and interoperability).

Standard risk assessment methodologies (fault-tree analyses, simulations, and red teams) have uncertain applicability and future analysis potential because information systems are very complex and threats can be very diabolical. Information security responsibilities are decentralized, and specific system vulnerabilities that are discovered are very sensitive and tightly controlled (for good reasons).

Undiscovered risks may continue to be the greatest concern. This suggests that continuing vigilance is required so that known problems can be fixed as they are discovered (if costs to fix them are "reasonable"). If known problems are hidden but not fixed, threats can be monitored and contingency plans developed, but associated risks may be impossible to measure in terms of direct (immediate) loss potential (such as human lives, repair and replacement costs, and opportunity costs while equipment is down).

With the above caveats hopefully lowering expectations as to the precision achievable, a preliminary assessment of the current state of first-generation SIW in terms of the key dimensions listed above is

1. **Number of offensive SIW players:** *Unknown* (but probably between 0 and a few).

2. **Tactical warning** (Is an attack under way?) and **attack assessment** (TW/KA) (If so, (by whom, how big, and at what?): The issues are uncertainty in perpetrator identity and the potential value and timeliness of warning indicators. All are *unknown,* but perpetrator uncertainties will likely be small in first-generation SIW in which IW is only one element of the conflict (but could be *large* if the perpetrator so desires).

3. **Damage assessment** (size and scope of damage): Significant damage will speak for itself; most critical damage assessment issues concern the potential for, and the implications of, further damage.

4. **Uncertainty in weapons effects:** Large.

5. **Degree of SIW vulnerability:** *Unknown* (but there are worrisome trends and real concerns).

Although we do not know with confidence what the current situation is concerning offensive and defensive SIW capabilities, people with informed *opinions* tend to fall into one of two groups: (1) those who see the historical glitches in information infrastructures as indicative of potential vulnerabilities that could be exploited by future adversaries, possibly with significant strategic advantage, and (2) those who see this past experience as strong evidence that the exploitable effects of whatever vulnerabilities might exist would be relatively modest and that the systems are evolving in a "Darwinian" mode that will continue to ensure appropriate defense mechanisms i.e., that there will never be such a thing as *strategic* information warfare. Determining the correct view between these two positions is less important than how we should proceed, given current (and future) uncertainties.

## ALTERNATIVE FIRST-GENERATION SIW END STATES

The fourth step in the SIW framework design process is the crafting of a set of *plausible and potentially desirable* alternative first-generation SIW asymptotic end states, taking into account the nature of the first-generation SIW threats that have been expressed in terms of the previously mentioned key dimensions of the first-generation SIW environment. Note the criterion "plausible and potentially desirable," which eliminates possible end states such as a very large number of nations with "major-league" offensive SIW capabilities alongside generally poor defensive SIW capabilities.

This end state crafting process is in effect likely to be an aggregation of assessments of the impact and possible future evolution (shaped or not shaped by related targeted strategy and policy decisions) of a set of threats identified in various SIW scenarios—expressed to the degree possible in terms of the key dimensions.

On the basis of the above approach, an initial array of possible alternative first-generation SIW asymptotic end states might be

- **A U.S. supremacy in offensive and defensive SIW.** The United States overwhelmingly dominates the SIW environment because it possesses

  — The world's best offensive SIW tools and techniques, capable of penetrating any other country's SIW defenses

  — Highly effective SIW defenses and reconstitution and recovery capabilities, which it *selectively* shares with allies, effectively reducing the vulnerability of potential SIW targets in the United States (such as key U.S. infrastructures) to strategically insignificant levels

  — Traceback capabilities that result in a very high level of confidence in perpetrator identification capabilities, whereas no other nation has traceback capabilities good enough to identify the United States as the source if it launches SIW attacks.

- **Club of SIW elites.** Through a combination of technical capability and resource allocation, an international group of highly competent SIW nations (5–10) emerges, with the United States almost certainly the most competent of the group. Mutual deterrence of SIW use is the common goal among club members. This handful of SIW "major leaguers" collaborates with each other to

  — constrain the spread of major-league SIW capability to other nations and non-nation actors

  — de-emphasize SIW and establish a norm of no first use of SIW

  — set international technical standards for cyberspace that help to perpetuate the exclusivity of the club.

- **Global "defense dominance" in SIW.** As a consequence of broad global cooperation in fielding very high quality SIW defenses, the vulnerability of key potential SIW targets (including key infrastructures) in most nations is reduced to strategically insignificant levels. This end state is further bolstered in some measure by international cooperation in the global dissemination of

— High-quality traceback capabilities (and/or a commitment to provide "whodunit?" traceback information in the event of a serious SIW attack).

— High-quality TW/AA) capabilities.

— Establishment of an SIW "arms control" regime along the lines of the BW and CW arms control regimes that establish international information operation norms, standards, legal restrictions, and enforcement mechanisms. Like currency counterfeiting, software piracy, and other threats to world economic order, SIW is something responsible states do not do. SIW rogues are dealt with as the U.N. dealt with Saddam Hussein: Deny them their goals and punish them.

- **Market-based diversity.** The extent of damage or disruption achievable in an SIW attack is modest, and reconstitution and recovery is fast because

— the natural strength of diversity in the globalization and standardization of cyberspace reduces overall vulnerability to SIW attack to moderate levels.

— global cooperation provides high-quality damage assessment tools.

— market-reinforced ("good neighbor") cooperation insures rapid reconstitution and recovery.

## ALTERNATIVE ACTION PLANS

The fifth step is applying the methodology to develop alternative action plans. The analytical and conceptual framework described here can be applied to concrete decisions affecting many areas of public policy. For government actions designed to address SIW vulnerabilities, the framework provides a step-by-step process of addressing the relationship between strategy and policy questions in the SIW domain, and the net, or relative impact of different policy choices on achieving overall SIW-related strategic objectives.

The process of developing a set of alternative action plans thus involves

1. choosing a set of illustrative alternative SIW end states

2. deciding on a selected set of key SIW strategy, policy, and related issues (such as those mentioned above), with an eye to moving in the direction of a specified end state.

Table S.2 provides a sample set of alternative action plans for navigating toward the four end states mentioned above. The plans are based on decisions on those SIW issues in the "Low-Hanging Fruit" and "Tough Issues" categories (see the Key Strategy and Policy Issues Section). Note that, in some instances, more than one alternative is compatible with the indicated end state. (A more detailed description of some of the more cryptic entries in Table S.2 is provided in the body of this report.)

## CONCLUSIONS

The strategy and policy decisionmaking framework and process, an evolving series of frameworks described above, appears to offer a useful means of organizing thinking about the emerging SIW problem and achieving an inaugural action plan in this arena. It should therefore contribute to the ongoing effort to identify those SIW-related issues on which decisions need to be made at this time in the United States, and the appropriate forum(s) in which to take up these issues.

This framework and process, though oriented to U.S. national decisionmaking, should also contribute to preparations for the imperative, and even more challenging, international decisionmaking process on this subject. The issue of the appropriate forum(s) for such an undertaking also remains to be resolved.

**Table S.2**

**Alternative Action Plans**

| Key Strategy and Policy Issues | Competition | Mixed (Competition and Cooperation) | Cooperation | |
|---|---|---|---|---|
| | A | B | C | D |
| | U.S. Supremacy in SIW | Club of SIW Elites | Global "Defense Dominance" in SIW | Market-Based Diversity |
| Locus of Responsibility/Authority | Federal government leads; national security focus<br>Joint leadership | Federal government leads; national security focus<br>Joint leadership | Federal government leads; law enforcement focus<br>Joint leadership | Industry leads |
| Tactical Warning and Alert Structure | Government-led NICON model<br>Counterterrorism model | Government-led NICON model<br>Counterterrorism model<br>CDC model | CDC model<br>Industry-led model | Industry-led model |
| Declaratory Policy (Links with other Military Instruments) | Strong retaliation threat (SIW retaliation emphasis)<br>Reassurance on invulnerability of key infrastructure | Moderate retaliation threat vs. nonclub actors<br>Some reassurance on invulnerability of club infrastructures | No retaliation threat<br>Reassurance on resilience of GII | Moderate retaliation threat (emphasis on economic instruments) |
| International Information Sharing and Cooperation | SIW programs compartmentalized | High degree of cooperation within club (G-7/FATF model) | High degree of cooperation<br>Institutional links through NATO, FATF, etc. | High degree of voluntary cooperation |
| Vulnerability Assessments | Government-led (NICON organizational model) | Government-led (G-7/FATF model) | Public/Private U.S. (WHO Model) | Public/Private U.S. (CDC Model) |
| R&D/Investment Strategy Priorities | National security-oriented protection goals<br>Some coordinated defensive R&D with Allies | Coordinated defensive R&D with Allies<br>Some proscriptions on offensive SIW R&D | Coordinate defensive R&D with allies<br>Proscriptions on offensive SIW R&D | Proscriptions on offensive SIW R&D<br>Private sector focus |

## WHAT IS STRATEGIC INFORMATION WARFARE?

### INTRODUCTION

"What is 'Strategic Information Warfare?'" was also the title for the first chapter in the initial RAND publication[1] on this emerging subject. The question bears repeating. Is there a useful political-military strategic concept, which can be called Strategic Information Warfare (SIW), that can be viewed as the intersection of strategic warfare and information warfare (see Figure 1.1)?

When considering this question and concept, what exactly does the term *strategic warfare* mean today, when compared with the past? What might the character be of

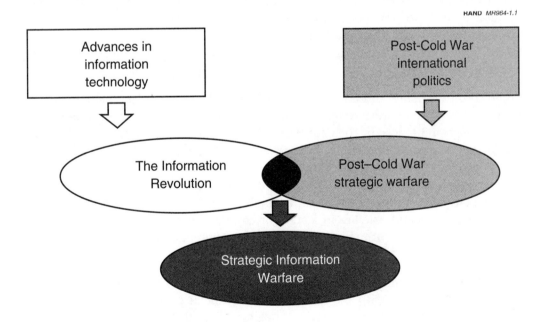

Figure 1.1—Strategic Information Warfare

---

[1]Roger C. Molander, A. S. Riddile, and Peter Wilson, *Strategic Information Warfare: A New Face of War*, Santa Monica, Calif.: RAND, MR-661-OSD, 1996.

strategic warfare in the future? What does the term *information warfare* mean today? What will it mean in the future? Is the concept of an intersection between strategic warfare and information warfare truly new—or has such a construct existed in the past?

A brief look at these and related questions follows, before the objective of this report, the development of a framework for facilitating near-term decisionmaking on SIW-related strategy and policy (and related) issues, is addressed.

## WHAT IS STRATEGIC WARFARE?

The term *strategic warfare* is time honored. *Webster's New-Collegiate Dictionary*[2] offers the following definitions of *strategic*: "of great importance to an integrated whole" and "striking at the sources of an enemy's military, economic, or political power."

The *Concise Oxford Dictionary*[3] definitions are also noteworthy: "essential in war" and "designed to disorganize the enemy's internal economy and destroy morale."

During the Cold War, strategic warfare came to be synonymous with nuclear warfare, at least in the United States and the Soviet Union, almost to the exclusion of other potential forms of strategic warfare. But the end of the Cold War came very fast and very unexpectedly. No one in the United States or the Soviet Union (or in other countries) had given much thought to what strategic warfare would be like in the absence of the Cold War. For example, no one had considered what the character of strategic warfare might be for a global power like the United States in a multipolar world where plausible U.S. adversaries might have regional rather than global strategic objectives.

Some countries (Israel and Vietnam are good examples) had, in fact, been forced to think about strategic warfare in regional terms for many years (even though they were strongly influenced by the Cold War). Furthermore, much of the United States' major experience in-20th-century warfare outside the Cold War (that is, World War I and World War II) was largely regional strategic in character (although global in the sum of its parts). Nevertheless, the United States had not given much thought to how post–Cold War regional adversaries might seek to gain strategic leverage over the United States and its allies in a crisis or conflict, much less to how they might achieve such leverage through means other than a direct confrontation of conventional forces.

Might future regional adversaries, especially because the Persian Gulf War made conventional conflict with the United States so clearly unappealing, be highly attracted to the search for asymmetric strategies? Might they implement such strategies by exploiting weapons of mass destruction, such as nuclear, chemical, or biological weapons? Or by the selective exploitation of highly advanced conventional

---

[2]Merriam-Webster Inc., *Webster's New Collegiate Dictionary*, Springfield, Mass., 1997.

[3]*The Concise Oxford Dictionary,* Oxford: Oxford University Press, 1982, p. 1052.

forces emerging from the "Revolution in Military Affairs"? Or by the exploitation of the Information Revolution to hold at risk key national strategic assets (see Figure 1.2)? All of these strategies would appear to be distinct possibilities.

At the same time, the United States could face what might be called "global peer competitors" who would incorporate SIW as part of a broad array of strategic weapons with which to confront the United States (see Figure 1.3). When one couples the question of what is going to be in future strategic weapon arsenals (including new kinds of economic weapons) to the highly dynamic and multipolar character of the international security environment, it becomes clear why the future of strategic warfare appears to be highly uncertain in terms of both (1) the strategic objectives of prospective adversaries, and (2) the potential means for exerting strategic leverage that might be at their disposal.

## WHAT IS INFORMATION WARFARE?

In contrast to strategic warfare, information warfare is a relatively new term that has found its way into the U.S. and international security lexicon only in the past few years, though the concept of the use of information in warfare is hardly new. The emergence of the term information warfare and its prominence can probably be directly tied to the Information Revolution, and to an expanding belief that this emerging revolution is so strong and potentially far-reaching that it could produce a new facet of modern warfare, or even a new kind of warfare.[4]

Through the mid-1990s, the term information warfare surged and then languished. It consistently defied clear definition, much less a consensus definition. Often, the term seemed too broad, encompassing traditional military areas, such as battlefield command and control warfare (C2W) and other traditional forms of electronic warfare (EW), that were evolving in response to the Information Revolution, but not necessarily changing dramatically. Some of these more traditional forms of warfare had, in fact, been "driving" the Information Revolution.

Something new in the general information warfare arena was, however, gaining increasing credence: the possibility that future adversaries might exploit the tools and techniques of the Information Revolution to hold at risk (not to destruction but to large-scale or massive disruption) key national strategic assets, such as initial elements of the national military posture or the national infrastructure sectors.

The utility and the applicability of the term information warfare as a broad rubric thus became increasingly a matter of debate. In response, the DoD developed a new lexicon and typology for the broad subject of "information operations" as a more appropriate broad rubric for this general subject area, while acknowledging that in actual conflict some aspects of information operations will befit the label information warfare, if not strategic information warfare.

---

[4]Definitions of information warfare vary. An example of scholarship's relating information warfare to more traditional military challenges is included in Martin Libicki, *What is Information Warfare?* Washington, D.C.: Center for Advanced Concepts and Technology/National Defense University, 1995.

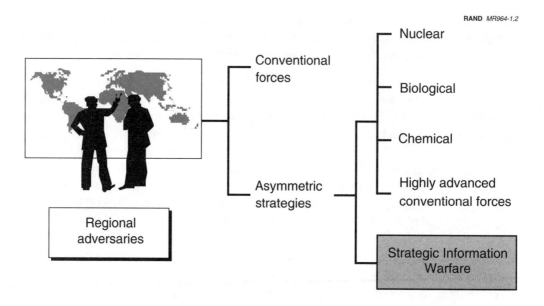

Figure 1.2—Asymmetric Strategies That Might Be Sought by Future
U.S. Regional Adversaries

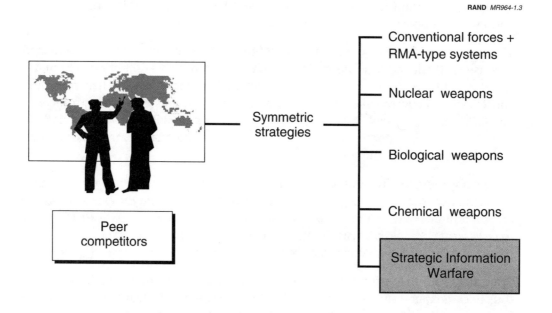

Figure 1.3—Symmetric Strategies That Might Be Presented by Future U.S. Peer Competitors

In this evolving definitional context, the use of the term strategic information warfare to describe what might be called the high end of the potential impact of the Information Revolution on warfare—a foreseeable intersection between strategic warfare and information warfare—seems appropriate.[5]  It was becoming clear that this kind of new strategic infrastructure threat could manifest itself in a future international strategic crisis or conflict involving the United States in two important ways[6, 7]:

1. A threat to U.S. national economic security. The holding at risk for massive disruption of key national infrastructure targets, to such a degree that a successful attack on one or more infrastructures could produce a strategically significant result (including public loss of confidence in the delivery of services from those infrastructures).

2. A threat against the U.S. national military strategy.  The possibility that a regional adversary might use SIW attacks to deter or disrupt U.S. power projection capability.  Concerns here include information warfare threats against infrastructure targets in the United States that are vital to overseas force deployment, and threats against comparable crucial infrastructure targets in allied countries.  A key regional ally or coalition member under such an attack might refuse to join a coalition—or worse, quit one in the middle of a war.

The following section discusses whether there are precedents for such a strategic warfare concept, and how it might evolve in the long term?

## THE HISTORY AND FUTURE OF STRATEGIC INFORMATION WARFARE

Has the concept of strategic information warfare, a strong information component of strategic warfare, existed in the past?  How important has it been?

Although strategic information warfare is a relatively new term, the concept of an information component of strategic warfare is not.  In fact, it may be hard to find any conflict worthy of the name strategic warfare that did not manifest some important information facet.  (Sun Tzu, for example, recommended the creative use of information to achieve strategic objectives while avoiding conflict.)  One could probably even note several historical instances in which fundamental changes in technology produced fundamental changes in the character of the information component of strategic warfare.

At the same time, the potential impact of the Information Revolution on strategic warfare may be unprecedented.  Whereas strategic information warfare may, in the past, have played largely a subordinate role—in early times in the strategic impact of, for example, conventional armies and navies and later, likes of airplanes, rockets, and/or nuclear weapons—it might play a much greater role in strategic warfare in the wake of the Information Revolution.

---

[5]It is clear from the media and the international literature that the use of the term *information warfare* has increased considerably in both the public and the international arena, albeit increasingly as a shorthand for what is here labeled strategic information warfare.

[6]Roger C. Molander, A. S. Riddile, and Peter Wilson, *Strategic Information Warfare: A New Face of War*, Santa Monica, Calif.: RAND, MR-661-OSD, 1996.

[7]Roger C. Molander and Peter Wilson, The *Day After...in the American Strategic Infrastructure*, Santa Monica, Calif.: RAND, MR-963-OSD, 1998.

Moreover, the potential impact of the Information Revolution on the vulnerability of key national infrastructures and other strategic assets may over time give rise to a brand-new kind of information-centric strategic warfare that is worthy of consideration independent of other potential facets of strategic warfare.

Therefore, it would appear (see Figure 1.4) that SIW as the future intersection between strategic warfare and the Information Revolution might be thought of in the following terms:

1. **First-Generation SIW.** SIW as one of several components of future strategic warfare, broadly conceptualized as being orchestrated through a number of strategic warfare instruments (see Figures 1.2 and 1.3).

2. **Second-Generation.** SIW as a freestanding, fundamentally new type of strategic warfare spawned by the Information Revolution, possibly being carried out in newly prominent strategic warfare arenas (for example, economic) and on time lines far longer (years versus days, weeks, or months) than those generally, or at least recently, ascribed to strategic warfare.[8]

As can be inferred from the above choice of terms, for established powers such as the United States, the authors tend to believe that first-generation SIW is more likely to be initially manifested. It is recognized, however, that is proposition is arguable. The United States, for example, might soon find itself in a situation in which it chose to exploit its current IT advantages and employ second-generation SIW, to prevail in a crisis that otherwise would have led to troop deployments and almost certainly to high numbers of casualties. See Appendix A for examples of a first-generation and a second-generation scenario.

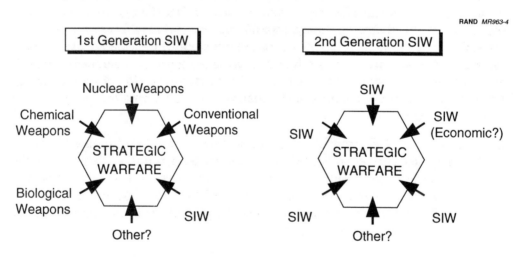

RAND *MR963-4*

Figure 1.4—Two Concepts of Strategic Information Warfare

---

[8]See John Arquilla and David Ronfeldt, *In Athena's Camp: Preparing for Conflict in the Information Age,* Santa Monica, Calif.: RAND, 1997.

For less-developed nations, which may not possess any other strategic weapons, second-generation SIW may be the first to appear. Second-generation SIW use by or against lesser powers might follow close on the heels of the demonstration of first-generation SIW.

Work to date has been able to identify a wide range of plausible examples of first-generation SIW. Most of these are rooted in the holding at risk for massive disruption of key infrastructures as part of a "combined arms" operation that includes the use of traditional military instruments of war. While the plausibility of such scenarios can be fairly well established, and there is great utility in their examination, it is far too early to discuss the probability of any such scenarios occurring according to any particular timetable.[9]

The authors found second-generation SIW scenarios more difficult to formulate. Not surprisingly, strategy and policy issues associated with such warfare are at this stage very difficult to conceptualize.

In light of this situation, this effort focused on the development of a decisionmaking framework for those problems associated with first-generation SIW concepts and their impact on established powers such as the United States (while also presenting at least one example of second-generation SIW in the next chapter).

Any substantial effort at this time to develop strategy and policy decisionmaking frameworks to address the types of problems manifest in second-generation SIW concepts for established powers such as the United States was viewed as premature, and best left to follow a more thorough examination of first-generation SIW concepts. There is, of course, a distinct possibility (if not the hope) that the approach to formulating first-generation SIW decisionmaking frameworks will prove to be highly useful in formulating comparable second-generation SIW frameworks.

---

[9]See Roger C. Molander, A. S. Riddile, and Peter Wilson, *Strategic Information Warfare: A New Face of War*, Santa Monica, Calif.: RAND, MR-661-OSD, 1996, and Roger Molander and Peter Wilson, The *Day After...in the American Strategic Infrastructure*, Santa Monica, Calif.: RAND, MR-963-OSD, 1998, for examples of two first-generation SIW exercise scenarios.

# Chapter Two

# THE STRATEGIC INFORMATION WARFARE
# FRAMEWORK PROBLEM

## THE NEED FOR NEW DECISIONMAKING FRAMEWORKS

What is a strategy and policy decisionmaking framework? Its most useful form is likely to be a series of relatively simple steps, or process, that presents the strategy and policy issues that need to be addressed in some particular arena in a logical architecture and along a logical path in a manner that facilitates decisionmaking on those issues.

New strategy and policy decisionmaking frameworks are born in the crucible of necessity (or perceived possible imminent necessity) manifested when a specific problem area (1) appears to demand action (or might soon demand action), and (2) is of such a nature that no readily applicable decisionmaking framework to forge an implementable action plan is available.

In some of these situations, an older decisionmaking framework may have been tested for its applicability to the needs of the problem area and found to be wanting. Use of such an older framework may, in fact, have been championed by those who favor formulating the problem area as a rapidly evolving old problem area rather than a new problem area. Failed attempts to apply an older decisionmaking framework may even have contributed to a delay in the more forthright expression of the need for a new framework.

During the gestation period that precedes the demand (or the proactive opportunity) to forge a new decisionmaking framework, the most urgent of the strategy and policy issues concerning the problem will presumably have begun to take shape. In fact, these evolving issues will undoubtedly have played a key role in catalyzing the perceived need for a new framework. These issues in effect lay in wait as crucial tests for the applicability of any proposed framework.

First-generation SIW as described in the previous chapter presents a set of issues that fit the above description of a new problem area in which the need for a viable initial strategy and policy decisionmaking framework appear to be clear. At the same time it cannot be claimed that there is a strong consensus demand yet for strategy and policy decisionmaking in this arena. Many argue that such a demand will not occur until there is an "electronic Pearl Harbor" in which massive disruption occurs in one or more key U.S. infrastructures or until some comparable event occurs in another nation. An additional problem is that no part of American society has clear responsibility in a "social contract" sense for this new type of problem.

These factors emphasize the challenge of pursuing a proactive approach to this problem area. It may be very hard to motivate the action—for both decisionmaking and implementation—that a close look at this subject appears to demand.

## AN EVOLVING SERIES OF FRAMEWORKS

As stated in the official project description, "The goal of this project is to formulate a common DoD strategy and policy framework for addressing the challenge of strategic information warfare." In carrying out this task, a single, temporally stable framework was initially sought to serve the stated function for SIW; however, it was soon concluded that the concept of *a single framework* to serve this function at this stage of the development was illusory. Rather, the correct construct for responding to a new strategic warfare component, one truly worthy of the label strategic as opposed to just another "strategic warfare wannabe," would have to be *dynamic*—in this case capable of responding to ongoing changes in both the international security and IT environments. The correct construct would have to be (1) *an evolving series of frameworks* that recognized and accepted the "punctuated equilibrium" realities of convening and executing strategy and policy decision making processes and (2) *a process* that recognizes and supports the dynamic and highly evolutionary character of such a construct (especially in its early stages). The need to conceptualize the process of responding to a potential new strategic threat in dynamic terms is therefore critical, especially at the outset.

This recognition can be seen as liberating, because the initial framework for strategy and policy analysis and decisionmaking can be seen as temporary, perhaps even as a short-term expediency. However, the concept of an evolving series of frameworks emphasizes the importance of making the basic components or dimensions in the *initial formulation* of the framework as internally consistent and broadly applicable as possible, even if there continues to be many unanswered questions about the future of both the technology involved and the international security environment.

A recent historical precedent for conceptualizing an evolving series of frameworks (a kind of "punctuated equilibrium") as the appropriate construct to aid analysis in and decisionmaking on new strategic threats, at least for first-generation SIW, can be found in the U.S. Cold War response to the Soviet strategic nuclear threat (see Appendix D).

In that response, the framework for U.S. decisionmaking went through (1) an "initial formulation" period right after World War II; (2) a major and broader reconceptualization of the problem in the late 1940s and early 1950s; (3) a transition period that lasted at least the following two decades, was marked by several major U.S. and Soviet decisionmaking junctures, and was punctuated by the Cuban Missile Crisis; and (4) a period of consolidation focused on working toward an asymptotic goal of strategic stability within an overall mutual deterrence concept. The latter period was of course interrupted and fundamentally altered by the collapse of the Soviet Union and the end of the Cold War.

One can anticipate that the framework component of the response to the first-generation SIW threat will follow a comparable historical pattern, perhaps including

(1) an initial formulation period in the late 1990s, (2) a transition period marked by the tracking of evolving information technologies and a dynamic international security environment (and an ever-evolving series of decisionmaking frameworks) that might last for more than a decade, and eventually (3) a period of consolidation when (and if) there is a consensus on an asymptotic goal for this type of SIW.

The possibility that a consensus asymptotic goal might emerge for first-generation SIW is discussed in detail in this report, both as an analytic tool to aid in the formulation of a suitable first-generation SIW decisionmaking framework and as an actual goal (akin to the eventual grudging settlement on mutual deterrence as the long-term goal in the Cold War).

One thought-providing way to possible long-term goal question in terms of first-generation SIW would be, "What is strategic information peace?"

The number of lessons, beyond the experience of the evolving series of frameworks, that can be "imported" from the Cold War is highly problematic. The problems, as this report clearly demonstrates, are *very* different from those of the Cold War. In addition, one can be even more skeptical about what the strategic nuclear warfare experience brings to the consideration of second-generation SIW.

## INITIAL FORMULATION OF A FIRST-GENERATION SIW STRATEGY AND POLICY DECISIONMAKING FRAMEWORK

A primary objective in conceptualizing the SIW decisionmaking framework problem is that the initial formulation of such a framework be one that can evolve in response to changes in its environment. It needs to have evolutionary potential rather than just being a temporary expedient that got decisionmaking going but did not have much utility thereafter.

The initial formulation also takes on increasing importance when one recognizes that it is like painting on blank canvas, and that, as the supporting framework for decisionmaking on an initial SIW action plan, it will draw special attention from all interested parties, including the media. Consequently, it will be seen as and will in fact be a painting done with four-inch brushes and primary colors by artists not yet certain of themselves.

The process of designing an associated inaugural first-generation SIW decisionmaking framework can be divided into the following distinct steps (see Figure 2.1):

1. Key dimensions of the SIW environment. Gain an understanding of the key dimensions of the future first-generation SIW environment or battlespace, that is, those dimensions of that environment that might in principle be shaped or influenced (presumably in some favorable direction) by effective near-term strategy and policy decisionmaking. Achieve this objective through (1) identifying the principal *defining features* of first-generation SIW within a spectrum of plausible first-generation SIW contexts, and (2) selecting those features that might be cast as *key dimensions* amenable to change as described above.

RAND *MR964-2.1*

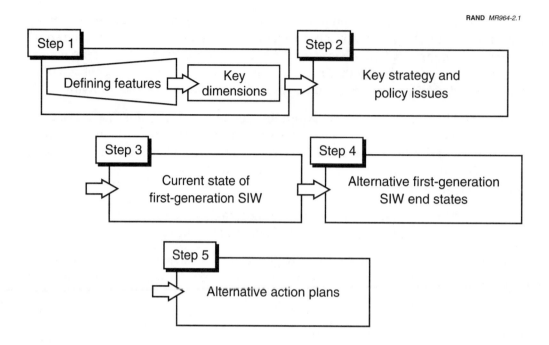

**Figure 2.1—Designing a First-Generation SIW Strategy and Policy Decisionmaking Framework**

2. Key strategy and policy Issues.  Identify those key strategy and policy issues (and other issues such as organizational issues) germane to the first-generation SIW problem (that is, issues on which near-term decisionmaking could shape the above-identified key dimensions of the SIW environment).

3. Current state of first-generation SIW.  Assess the current state of first-generation SIW in terms of absolute and relative offensive and defensive SIW capabilities, to better understand where one starts from today in the SIW realm.

4. Alternative first-generation SIW "end states."  In light of the above-mentioned first-generation SIW contexts and scenarios, craft a set of (plausible and potentially desirable) alternative first-generation SIW "end states" that are expressed in terms of the above-mentioned key dimensions of the first-generation SIW environment.

5. Alternative Action Plans.  Array the key SIW strategy and policy issues against each of these alternative end states, and conceptualize alternative action plans for moving toward one or more of these end states.

In pursuing the above approach, one must keep in mind that, at any one time, there will be limits to how far one can see into the "foreseeable" future.  This is especially true in the above articulation of alternative SIW end states in which the current dynamic character of both the IT and international security environments makes it inherently difficult to accurately characterize the real eventual alternative end states for first-generation SIW (much less for the broader second-generation SIW concept).

As noted previously, the true test for the utility of any decisionmaking framework, such as that shown in Figure 2.1, will be whether it proves it to be a sound basis in practice for achieving and implementing an initial set of first-generation SIW strategy and policy decisions. As will be clear later in this report, in this situation a further test is whether the framework will support the unprecedented and unparalleled international cooperation that may be necessary to navigate successfully toward a chosen end state.

Any such framework will have to be continually tested and evaluated against emerging contingencies (for example, through SIW exercises). One should recognize, however, that it may be hard to achieve a sustained high level of comfort concerning the viability of any such framework until the related IT and international security environments are less dynamic.

Each of the above SIW framework design steps is discussed in more detail in light of these considerations in the following chapters.

# Chapter Three

# KEY DIMENSIONS OF THE SIW ENVIRONMENT

## FROM DEFINING FEATURE TO KEY DIMENSIONS

The principal sources of the defining features of the first-generation SIW problem as presented in this chapter are (1) various RAND projects[1] (2) the SIW scenario vignettes described in Chapter Two, and (3) the final report of the President's Commission on Critical Infrastructure Protection issued in October 1997. Other sources have included a RAND-Ditchley conference on this general subject[2].

Table 3.1 presents the defining features and the basic consequences of these defining features in the real world of potential SIW conflict, and then recasts the defining features of Table 3.1 as key dimensions of the SIW environment that in principle can be both measured and changed by asking the following question: Is this defining feature of SIW a largely inaccessible characteristic of the First-Generation SIW environment, or *can it be potentially shaped or influenced in some favorable direction* by effective near-term strategy, policy, and related decisionmaking?

It is important to acknowledge the limited extent of SIW scenario space explored so far and the possibility that this might produce some bias in characterizing the threat (including its nature, magnitude, and timing) and deriving conclusions about possible responses. Because there have been no known real cases of SIW as defined here, considerable uncertainty exists about the timetable for threat emergence and the possibility that certain postulated vulnerabilities will emerge.

Nevertheless, we believe that the "Lewis and Clark–type" of exploration of the SIW landscape reflected in the above-mentioned experience is adequate to justify the rough level of structure and substance in the presentations in Table 3.1 and the assessment in this report that there are SIW-related issues that require urgent national and international attention.

---

[1]Roger C. Molander, A. S. Riddile, and Peter Wilson, *Strategic Information Warfare: A New Face of War,* Santa Monica, Calif.: RAND, MR-661-OSD, 1996; Robert H. Anderson and Anthony C. Hearn, *An Exploration of Cyberspace Security R&D Investment Strategies for DARPA: The Day After in Cyberspace II,* Santa Monica, Calif.: RAND; MR-797-DARPA, 1996; Roger C. Molander and Peter Wilson, *The Day After...in the American Strategic Infrastructure,* Santa Monica, Calif.: RAND, MR-963-OSD, 1998; and Roger C. Molander, David Mussington, and Peter Wilson, *Cyberpayments and Money Laundering: Problems and Promise,* Santa Monica, Calif.: RAND, forthcoming;

[2]Richard Hundley, et al., *Security in Cyberspace, Challenges for Society: Proceedings of an International Conference,* Santa Monica, Calif.: RAND, CF-128-RC, 1996.

**Table 3.1**

**Defining Features, Consequences, and Key Dimensions of the SIW Environment**

| Defining Features | Consequences | Key Dimensions |
|---|---|---|
| Entry cost low | May be many actors in the SIW battlespace | Number of offensive SIW players |
| Strategic intelligence on threat unavailable | Identity and capabilities of potential adversaries may be unclear | Number of offensive SIW players |
| Tactical warning difficult | May not know attack is under way | Tactical warning capability |
| Attack assessment difficult | May not know perpetrator or targets | Attack assessment capability, including perpetrator identity |
| Damage assessment difficult | May not know full implications of the attack | Damage assessment capability |
| Traditional boundaries blurred | May not know who has various responsibilities before, during, or after attack | N/A |
| Weapon effects uncertain | Both attacker and defender may be uncertain about weapon effects | Uncertainty in weapon effects |
| Infrastructure vulnerabilities uncertain but suspect | U.S. homeland may not be a sanctuary; vulnerable partners could make sustaining coalitions more difficult | Degree of SIW vulnerability |

# Chapter Four

# KEY STRATEGY AND POLICY ISSUES

## THE ISSUE MENU

This chapter identifies those key SIW-related strategy, and policy, and related issues that are, or soon will be, candidates for U.S. (and potentially multinational and international) decisionmaking.

The principle sources for the assessment of key SIW strategy, policy, and related concerns presented here are the same as those cited for the key features of SIW in the previous chapter. In particular, in various RAND "Day After..." exercises, the final step challenges participants to identify those key strategy, policy, and related issues germane to the first-generation SIW problem.

Based on these sources, the following SIW strategy, policy, and related issues appear to be candidates for decisionmaking:

1. Locus of responsibility and authority. Who should have the lead responsibility in the U.S. national response to the SIW threat—government (and, if so, which entities within the government?) and/or industry (and, if so, who within the key infrastructures)?

2. Tactical warning, attack assessment, and emergency response. How should governments and industry in United States (and throughout the world) organize to develop and implement capabilities and procedures to sense and respond to SIW threats?

3. Vulnerability assessments. How should government and industry cooperate in undertaking a vulnerability assessment of key U.S. national infrastructures?

4. Declaratory policy on SIW use. Should the U.S. government declare its policy on the use of SIW and the relationship between the use of SIW and other strategic military and economic instruments?[1]

5. International information sharing and cooperation. What principles should guide international collaboration (in particular with allies and coalition partners) in the SIW domain? Is there an SIW parallel to extended deterrence? To extended defense?

---

[1]See The Defense Science Board Task Force on Information Warfare–Defense (IW-D), *Final Report*, November 1996.

6. R&D and investment strategy. What investment strategy should the United States pursue for (1) monitoring, identification, and traceback techniques, (2) attack assessment techniques, (3) defense and reconstitution techniques, and (4) damage assessment techniques?

7. Intragovernment and intergovernmental cooperation on politically sensitive privacy issues. Are new approaches to intelligence community (IC) and law-enforcement cooperation necessary to respond to the SIW threat while still meeting constitutionally based privacy concerns?

8. Minimum essential information infrastructure (MEII). Should the U.S. seek to facilitate the development of information infrastructure-related systems and procedures to insure minimum levels of infrastructure functionality during periods of disruption? Should the United States do this?

   • only to serve well-defined government missions such as insuring the timeliness of military power projection plans?

   • to ensure minimal functional service level in key infrastructure sectors for U.S. society as a whole?

9. Encryption Policy. Should the United States mandate some level of encryption protection for information-dependent infrastructures?

In considering the breadth and character of the above issues and the challenges thus presented to existing decisionmaking processes, some sequencing in taking up these issues is needed nationally and internationally. To this end, these could be categorized issues as follows:

• Low-hanging fruit. Those issues that could be moved to closure nationally (or internationally) without undue difficulty, after suitable processes are identified or established.

• Tough issues to be faced now. Urgent but contentious strategy, policy, and other issues related to the inaugural charting of long-term SIW-related national goals and strategy.

• Deferred issues. Issues that, for one reason or another (for example, technical uncertainties), are not yet ready to be taken to closure—or, worse, issues that might produce "bad" strategy or policy decisions that would be hard to undo, if they were taken to closure prematurely.

Of the nine issues listed above, the first six, in whole or in part, appear to be either low-hanging fruit or inescapably urgent near-term issues. However, for issue 7 (intragovernmental and intergovernmental cooperation on politically sensitive policy issues), although this needs to be included in any discussion of SIW, more detail is needed on how privacy rights would be protected under specific strategies and policies.

For issue 8 (MEII), more analytical and conceptual work is needed to determine whether the MEII concept is feasible from both a technical and cost standpoint. For issue 9 (encryption policy), SIW is just one of many issues that need to be considered when the United States and the international community chart long-term encryption-related goals and strategies.

The first six issues are discussed below. Emphasis is given to the challenge of crafting clear issue statements, and associated alternative courses of action, in a manner suitable for decisionmaking.

## LOCUS OF RESPONSIBILITY AND AUTHORITY

The nature of the SIW problem is so broad and far-reaching that no one part of U.S. or international society clearly has the "social contract" responsibility for key elements of the problem, much less leadership and organizational responsibility for the problem as a whole. This makes *getting started* on forging and implementing a national strategy on this problem extremely difficult.

Although the U.S. defense and intelligence communities have institutional mandates that give them responsibility for offensive SIW, this is not the case for defensive SIW. Without a national decision on responsibility and authority in this arena, progress on the problem will assuredly be slow.

What division of responsibility and authority between government and the private sector makes sense? How should organization (if any) at the international level take place? Five alternative near-term courses of action do appear to warrant consideration:

1. Federal government leadership with a national security focus

2. Federal government leadership with an law-enforcement focus (for example, Department of Justice [DoJ] leadership)

3. Joint international government leadership with a national security focus

4. Joint international government leadership with a law-enforcement focus

5. International industry leadership with government support.

## TACTICAL WARNING, ATTACK ASSESSMENT, AND EMERGENCY RESPONSE

It is increasingly clear that a critical need exists for the United States to develop and deploy far more effective means of SIW tactical warning and attack assessment, and to develop procedures for emergency response to SIW attack. While this need has been virtually unquestioned for several years, no clear delineation of responsibilities in this arena between government and industry has been made yet. The same holds true for the potential development of a global approach to these issues.

Many of the concerns raised by this issue are echoed in traditional National Security and Emergency Planning (NS/EP) thinking. Under this rubric, government agencies in general, and DoD in particular, share clearly defined responsibility for protection of critical national infrastructures. For privately owned infrastructures, government mandates implemented through licensing requirements and/or information gathering or reporting guidelines constitute the critical "other half" of the effectiveness response to infrastructure vulnerability.

Alternative models of tactical warning, attack assessment, and emergency response in infrastructure protection scenarios focus attention on the critical differences that exist between perspectives on infrastructure protection. Different approaches emphasize varying views toward public sector and private sector roles, institutional responsibilities within the public and private sectors, and divergent perspectives on the importance of international collaboration for infrastructure protection.

Four distinct models of tactical warning, attack assessment, and emergency response emerged from our analysis of available alternatives:

1. A government-led national security-oriented model (a NICON model[2])

2. A government-led law-enforcement-oriented model (a counterterrorism model)

3. A CDC model

4. An industry-led model.

Each of these models offers a distinct perspective on the nature of the infrastructure protection problem, and would guide decisionmaking on institutional and policy matters pertaining to this problem in slightly divergent directions. It is nonetheless true that the models of infrastructure protection warning and alert response are not mutually exclusive approaches. Rather, the different models propose different modalities in responding to the warning, assessment, and response challenge in infrastructure protection.

## VULNERABILITY ASSESSMENTS

The argument for an urgent vulnerability assessment for key U.S. and allied infrastructures has been made previously.[3] As with the tactical warning, attack assessment, and emergency response issue, the basic question is not so much whether a vulnerability assessment needs to be made, but instead who will do it. There is also the question of whether and how to undertake a broader, internationally based infrastructure vulnerability assessment in light of the dependence of both the U.S. military and U.S. international business on the viability of U.S. infrastructures.[4] The basic alternatives are

---

[2]Roger C. Molander and Peter Wilson, *The Day After...in the American Strategic Infrastructure*, Santa Monica, Calif.: RAND, MR-963-OSD, 1998.

[3]See Neumann, Peter G., *Computer-Related Risks*, New York: Addison-Wesley, 1995, for a list of vulnerabilities related to advanced computing and communications systems.

[4]It is worth noting at this point that two recently concluded projects relate directly to this subject. As noted earlier, the IC completed a classified assessment of the information warfare threat in mid-1997. In October 1997, the President's Commission on Critical Infrastructure Protection (PCCIP) completed a wide-ranging examination of the vulnerabilities of critical infrastructures to disruption by hackers, criminals, or terrorists. One can infer from the results of these activities that a high degree of uncertainty exists

1. A U.S. government–led and DoD assessment of U.S. vulnerabilities

2. A joint public and private sector effort by the United States and other key nations (such as G-7 and/or potential SIW peer competitors)

3. An international public-private partnership similar to that of the CDC or WHO

4. A joint U.S. industry–led and government-assisted vulnerability assessment.

A number of key distinctions exists among these alternatives, reflecting the relative newness of the subject, the challenge of getting started, and different concepts of what end state (see Chapter Six) should be navigated toward.

## DECLARATORY POLICY

Strong interest exists about how SIW will be incorporated into broader thinking about the future conduct of strategic warfare. There are questions about what kind of policy statement, if any, the United States (possibly in concert with other like-minded nations) might make on this topic, in particular, about how the United States might respond to a genuine and effective SIW attack. The basic alternatives are declaratory policy statements threatening

- retaliation principally in kind for such an attack

- retaliation principally by non-SIW military means in response to such an attack

- retaliation by economic means, including possibly economically oriented SIW means, in response to such an attack

- complete ambiguity about how the United States would respond to such an attack.

## INTERNATIONAL INFORMATION SHARING AND COOPERATION

International information sharing and cooperation is a critical issue that demands serious thinking about long-term national and global goals, which are addressed in Chapter Seven as asymptotic end states. In terms of SIW-related sharing (for example, of defensive or traceback techniques), the strategic sharing experience of the Cold War (for example, with Japan, South Korea, and NATO allies) is not applicable because of the fundamental differences between the SIW problem and strategic nuclear warfare. The most striking question concerns sharing defenses because of the close coupling in SIW (see Chapter Five) between offensive and defensive SIW tools, techniques, and strategies. Another issue in this area is that virtually all U.S knowledge of offensive SIW techniques is highly classified and compartmentalized.

The basic policy alternatives regarding sharing and cooperation are

- No significant international SIW cooperation

---

regarding the vulnerabilities of both individual sectors and the nation as a whole. For details of the PCCIP's findings, see its Web site at http://www.pccip.gov.

- Limited international cooperation focused on defensive techniques (such as the G-7 model)

- Broad international cooperation organized through existing multinational security arrangements (such as the NATO model)

- Broad international cooperation organized through global arrangements (for example, the WHO model)

- Broad voluntary international cooperation (such as international Computer Emergency Response Teams [CERTs]).[5]

## INVESTMENT STRATEGY

It is difficult at present to know what the most likely direction is for SIW strategy and policy development—and for attendant problems concerning decisions on R&D and investment strategy.[6] Will strong offensive SIW capability be a key component of long-term U.S. SIW strategy? Will there be international cooperation on defenses? More definitive information is needed on these and other key technical issues that will be obtained only through R&D investment.

Basic alternative courses of action for R&D investment strategy, based on varying priorities, are

- *National security–oriented network protection goals.* National security–oriented network protection goals drive the research agenda. Rank-ordered priorities for SIW-related R&D include defense-centric network protection goals, information assurance, monitoring and threat detection, protection and mitigation, vulnerability assessment and analysis, risk management and decision support, and contingency planning, incident response, and recovery. The public sector would directly underwrite R&D activities in each of these areas, with further incentives used to encourage research by the private sector in specific areas.

- *Coordinated defensive R&D with allies.* Joint R&D conducted with allies would emphasize the same overall priorities of R&D listed above. Public funding of basic research and the development of technology by private sector characterize this approach.

- *International proscriptions of offensive SIW technological R&D.* Defensive research priorities for this approach are, in rank order, monitoring and detection, protection and mitigation, information assurance, risk management, contin-

---

[5]Responses to infrastructure disruptions caused by natural disasters also may foster international collaboration on responses. See National Research Council, Computer Science and Telecommunications Board, *Computing and Communications in the Extreme: Research for Crisis Management and Other Applications,* Washington, D.C.: National Academy Press, 1996.

[6]It is worth noting that infrastructure protection is only one of many competing factors influencing the development of the Global Information Infrastructure. A report that captures many of the other important concerns of infrastructure stakeholders is Computer Science and Telecommunications Board of the U.S. National Research Council, *The Unpredictable Certainty: Information Infrastructure Through 2000,* Washington, D.C.: National Academy Press, 1996).

gency planning, incident response and recovery, vulnerability assessment, and threat detection.  Arms control–type restrictions may be imposed to proscribe particular types of R&D seem as unique or particularly applicable to offensive SIW development.  Private sector development of technologies would be encouraged, with basic research undertaken by public institutions.

- *Private sector focus.*  This approach involves informal consultations between government and industry on information infrastructure security and vulnerability issues.  R&D priorities would be determined by the private sector, but might resemble the following rank-ordered list:

  — Risk management
  — Contingency planning
  — Public sector research on technical standards
  — Public sector coordination of monitoring and threat detection
  — Activities conducted by the private sector.

# Chapter Five

# CURRENT STATE OF FIRST-GENERATION SIW

## ASSESSING AN EMBRYONIC CONCEPT

The objective of this chapter is to provide information about where the United States is starting from in terms of relative and absolute SIW capabilities. A macro assessment of the current state of first-generation SIW in terms of relative and absolute offensive and defensive SIW capabilities of the United States and other nations (or other parties) would be difficult to do at present, even at a classified level. The current dynamic character of the Information Revolution and the embryonic character of SIW as a potential political-military instrument both argue for caution in making such an assessment—classified or unclassified, at present and for the foreseeable future. Any such assessment at this stage (including that provided here) should therefore be viewed as a poorly focused and underexposed "snapshot" with shapes and colors for less distinct than one would prefer. In addition, any such assessment will inescapably blur over time and thus become less useful, with a decay that can likely be measured in multiple months rather than years.

It can therefore be anticipated that further international appraisal of the SIW problem in the near-term will lead to strategic defense responses in many nations and infrastructures in a "get your SIW defenses up" mode, with a strong orientation toward defending against potential U.S. intrusion. These circumstances will create a strong, international market for robust SIW defenses that could over time, or even overnight, negate some of the most effective offensive SIW tools and techniques in the United States or elsewhere.

Currently, the principal SIW assessment issues from a U.S. perspective are

- The extent to which hostile SIW powers already exist and the degree to which they can seriously harm the United States with SIW attacks

- The extent of current U.S. offensive SIW capability vis-à-vis other countries (foe, neutral, or friendly), whether overt or covert, in preventive, preemptive, or retaliatory SIW actions.

With the above caveats in mind, this chapter provides a rough, unclassified assessment of where the SIW area stands today.

## KEY FACTORS IN SIW DEVELOPMENT TO DATE

Although SIW is currently an embryonic concept, involving considerable uncertainties, it is possible to identify some of the evolutionary processes and events that have determined its present state. An assessment of some of those formative factors and experiences and of their implications follows.

First, as mentioned previously, SIW is a natural and logical Information Age extension of "information operations" carried out from the dawn of recorded history to deceive, demoralize, and disrupt the enemy. The means have changed, but the objectives are rooted in basic dimensions of political and military conflict. Today, however, the ubiquitous nature of information systems and the increasing dependence of societies and militaries on them suggest that the relative significance of information vis-à-vis other instruments of warfare—and thus of SIW—may be increasing. Can we find support for this hypothesis?

It is important to keep in mind that offensive and defensive SIW strategies and techniques are, to a large degree, opposite sides of the same coin. Effective defensive SIW capabilities cannot be developed without understanding offensive threats and their potential impact; offensive developments require an understanding of potential system and operational vulnerabilities, including defense limitations. Offensive and defensive SIW actions, therefore, are in large measure distinguishable only by intent.

The dual nature of offensive and defensive SIW, and the inherent need to protect potentially fragile capabilities, suggest that there may always be significant underlying uncertainties in this area. Although it is a truism that "we do not know what we do not know," in the area of SIW that may be the defining feature. We would therefore expect that early attempts such as this one to formulate an SIW framework must accept a certain level of *ambiguity* and *secrecy* concerning some of the key characteristics of SIW. Unlike the strategic nuclear warfare threats, which are both obvious and awesome in their general features, SIW may be of uncertain potential and actual impact for a long time.

For the above-mentioned reasons, there are fundamental and complex issues associated with any assessment of the current state (or any possible future state) of SIW. Offensive assessments imply defensive flaws, which, once recognized might be corrected, thereby reducing offense effectiveness. Furthermore, there may also tend to be more of a political stigma associated with offensive capabilities than with defensive capabilities—even though both have a legitimate and potentially essential role in any conflict. However, even though defensive capabilities might be more politically benign, they respond to implicit or explicit threats that might be able to respond to and defeat known defensive measures. This offensive-defensive threat dynamic suggests that one should be careful about how one characterizes and identifies SIW specifics. One would in general expect that any nation that has well-developed information technologies and applications will be both a target for, and a threat to, other information-oriented nations.

## ASSESSING CURRENT LEVELS OF OFFENSIVE SIW CAPABILITY

The United States as the global leader in the development and exploitation of information systems, has the potential to be an offensive SIW "superpower" if any nation does. Any lesser assessment of U.S. SIW potential compared with that of others would be judged as laughable by those nations that are beginning to speculate about the significance that SIW might have in future conflicts. But how far has this U.S. potential been exploited? How fast could it be exploited if the United States were to make a strong national commitment to the urgent development of offensive SIW capabilities?

On the offensive side, the current U.S. view of and experience with information operations is as a supporting but relatively low-profile element of U.S. military strategy and doctrine. The United States has well-developed and successful offensive C2W, EW, and other IW capabilities (for example, SOCOM is a master of psyops, and the military services develop and operate electronic warfare systems, as manifested in the large-scale use of C2W and the suppression of enemy air defenses in the Persian Gulf War), but these could hardly be characterized as "strategic" in the sense of this report. Offensive first-generation SIW, which by definition has the potential to "hold at risk" a country's central nervous system, is a much more sensitive undertaking than are "information operations" as supporting missions in conventional warfare. It is one thing to target military leadership, communications, and radar; it is quite another to target public utilities that, among other things, provide power to hospitals.

The sensitivities of our friends and allies and the political-military capital that might accrue to possible adversaries from increasingly open emphasis on U.S. offensive SIW initiatives has largely kept the more definitive information on these capabilities in the back room. Some SIW offensive capability clearly exists (and some has been demonstrated), but the full potential is politically and militarily sensitive. In the next chapter we will discuss the future significance of offensive SIW to the United States, but this chapter focuses on what other nations might be able to do to the United States.

To guide thinking about the offensive SIW goals that might have been set in the past and that are reflected in the current development of U.S. offensive SIW, it is appropriate to ask what offensive SIW capabilities an information superpower (and/or possible aspirant to SIW supremacy) might seek to develop and exploit at this embryonic stage of SIW development. The range of potential goals might include

1. Global SIW supremacy: The ability to hold at risk to massive disruption the key information infrastructures in all industrialized nations.

2. SIW strategic leverage: The ability to use SIW as an adjunct to leveraging conventional power projection capabilities, to better meet U.S. regional strategic military objectives.

3. Limited SIW potential: The ability to use effectively selected advanced and traditional (even "low tech") information operations (such as psyops) against less

technologically advanced opponents for strategic effect, but with no effort to develop offensive capabilities against peer competitors.

U.S. pursuit of the first of these goals would be consistent with either of the following perspectives: (1) casting SIW as a military instrument of last resort (akin to nuclear weapons), for example, as a "more humane" way of punishing rogue states, or (2) seeking to establish a global political-military environment in which U.S. SIW capability gives the United States a new and highly effective source of global leverage. Discounting the likelihood of the United States having already made SIW-related decisions based on the latter objective, a prior decision to posture U.S. SIW capabilities primarily as a deterrent means that the offensive capabilities must be well publicized and credible to possible opponents. This has certainly not so far been the case for the United States, much less for any other nation with advanced IT capabilities.

The second level of capability listed above represents a more modest goal, and is probably more plausible as a goal the United States might have set for itself in the past—and potentially could be in place today. This type of goal suggests that the information component of modern warfare is seen as becoming increasingly decisive at the strategic level, but so far only when coupled with more traditional military force elements and tactics.

The third level of capability reflects the asymmetries that might increasingly dominate conflict in the future—the "haves" versus the "have nots," "First World" versus "Third World," and so forth—will also characterize the SIW environment. It presumes that vulnerability to information operations, at least in a "strategic" sense, may be country-specific. Actually, less developed countries with less information and information-dependent infrastructures may have an inherent robustness to SIW capabilities that could bring more complex and developed nations to their knees. Thus, countries that are in the early stages of exploiting IT will likely initially move from a position of relative invulnerability to some level of vulnerability.

Which of these goals is reflected in any current U.S. offensive SIW capability—if any—seems likely to remain as much a matter of speculation as the actual level of capability the United States currently achieves.

## ASSESSING CURRENT LEVELS OF DEFENSIVE SIW CAPABILITY

While being a leading contender for augmenting its arsenal with offensive SIW capabilities, the United States, by virtue of its position in the world, is also a natural target for SIW attack. The Unites States leads the world in the development and application of IT and has a complex society and economy that are critically dependent on information systems. It is geographically protected and has the world's most awesome conventional military capabilities. If the United States is to be defeated militarily in the near future, it will most likely be because an enemy successfully uses an "asymmetric" strategy to achieve some strategic end.

As discussed in Chapter One, there are two general classes of IW threats that could become strategic: (1) threats to U.S. national economic security, and (2) more direct threats against the U.S. national military strategy. The enemy's tactics could range from threats to action, with results ranging from psychological to physical, at scales ranging from trivial to shocking, over time frames from days to years. (Appendix C provides a list of exemplary SIW weapons).

The first logical step in understanding SIW defensive implications is to conduct a review of potential U.S. vulnerabilities to conceivable SIW attacks across a broad spectrum of threats and scenarios. Unfortunately (or fortunately), we have very little "real-world" experience on which to base this assessment. A number of natural events (such as storms, and earthquakes), human errors (software and control), and purposeful mischief (hobbyist hackers and criminals) suggest that things can go wrong in various national infrastructures, occasionally on an impressive scale. However, none of these past events has been "strategic" in its impact nor do any appear to have been strategic in their intent.

One obvious problem with this paucity of defensive SIW-related experience is in relating cause and effect: Have we escaped SIW attacks because undetected attempts were not successful or because no one has made any such attempts yet?

Although a great deal of uncertainty surrounds the future vulnerability of information infrastructures, a number of trends seem to point toward an expanded dependence on inherently less secure networking concepts.[1] In particular, the widespread adoption of open network standards and technologies means that the industries and applications delivered over the GII may become more vulnerable to single-point failures. The growth of electronic commerce, the prospective expansion of electronic stored value (cyberpayment) payment systems, and plans for the delivery of critical services (such as telemedicine and government communications) over the GII all present potential attractive targets for an SIW attack.

Open network standards place a premium on high bandwidth reliability and interconnectedness in system operation. The delivery of products and services in an accessible manner to consumers means that security protocols are an important, but not principal, driver of network designs. The expansion of the GII offers potentially "frictionless" economic transactions, resulting in an increase in economic efficiency. At the same time, however, the tradeoffs between network access and greater security and authentication in network operation may yield a set of infrastructures that is less robust to information-based attack. Infrastructure protection programs may yield solutions to network vulnerability that violate the basic business models of many private sector infrastructure operators and users. Because most of the infrastructures are privately held, potentially troubling implications exist for the actual implementation of any agreed-upon set of network security recommendations. In short, the technical availability of defenses does not transmit simply and automatically into the real-world deployment of those systems to the GII.

---

[1]For an examination of the new features of network dependence, see William J. Drake, ed., *The New Information Infrastructure: Strategies for U.S. Policy*, New York: The Twentieth-Century Fund Press, 1995.

The defensive SIW assessment thus comes down to an assessment of information infrastructure vulnerability, threat potential, and vulnerability consequences. These assessments also have their problems. Existing information infrastructure systems are complex, dynamic, flexible, and interdependent. They are public and private, military and commercial. Some (such as those in the banking industry) have been "hardened" by design because the risks were obvious. Others have evolved in a more benign environment with functions not driven by threats (such as cost, accessibility, and interoperability).

Standard risk assessment methodologies (fault-tree analyses, simulations, and red teams) have uncertain applicability and future analysis potential because the information systems are very complex and the threats are very diabolical. Information security responsibilities are decentralized, and specific system vulnerabilities that are discovered are very sensitive and tightly held (for a number of very good reasons).

Undiscovered risks may continue to be the greatest ones. This suggests that continuing vigilance is required, so that known problems can be fixed as they are discovered (if the costs to fix are reasonable). If known problems are hidden but not fixed, threats can be monitored and contingency plans developed, but associated risks may be impossible to measure in terms of direct (immediate) loss potential (such as human lives, repair and replacement costs, and opportunity costs while equipment is down.).

## A PRELIMINARY ASSESSMENT OF WHERE WE ARE[2]

Keeping in mind the above-mentioned caveats concerning the precision achievable in assessing first-generation SIW, a preliminary assessment of its current state, in terms of the key dimensions discussed in Chapter Three, is as follows:

1. **Number of offensive SIW players**: *Unknown* (but probably between 0 and a few).

2. **Tactical warning** (Is an attack under way?) and **attack assessment** (If so, by whom, how big and at what?): The issues are uncertainty in perpetrator identity and the potential value and timeliness of warning indicators. All are *unknown*, but perpetrator uncertainty will likely be small in first-generation SIW in which IW is only one element of the conflict (but uncertainty could be *large* if the perpetrator so desires).

3. **Damage assessment** (size and scope of damage): Significant damage will speak for itself; most critical damage assessment issues concern the potential for, and the implications of, further damage.

4. **Uncertainty in weapons effects**: Large.

5. **Degree of SIW vulnerability**: *Unknown* (but there are worrisome trends and real concerns).

---

[2]This assessment is based on the recently completed work of the PCCIP. The commission's conclusions are available at its web site (**http://www.pccip.gov**). One can emphasize that the pronounced levels of uncertainty in the future of cyberspace mean that all statements of risk and vulnerability are subject to some degree of controversy.

Although we do not know with confidence where the United States stands now (and this may never really be known in the future), people with informed *opinions* tend to fall into one of two polar groups: (1) those who see the historical glitches in the information infrastructures as indicative of potential vulnerabilities that could be exploited by future adversaries, possibly with significant strategic advantage, and (2) those who see this past experience as strong evidence that the exploitable effects of whatever vulnerabilities might exist would be relatively modest and that the systems are evolving in a "Darwinian" mode that will continue to ensure appropriate defense mechanisms—there is no such thing as "SIW." Who is right at present is less important than how the United States and the world should proceed, given the current (and likely future) uncertainties. That issue is discussed in the next chapter.

# Chapter Six

# ALTERNATIVE FIRST-GENERATION SIW END STATES

## INTRODUCTION

As discussed in Chapter Two, the fourth step in the SIW framework design process is the crafting of a plausible and potentially desirable set of alternative first-generation SIW asymptotic end states. This process need to take into account the nature of the first-generation SIW threats that have been identified in terms of the previously mentioned key dimensions of the first-generation SIW environment. The "plausible and potentially desirable" criteria eliminates such possible end states as a very large numbers of nation with "major-league" offensive SIW capabilities but poor defensive SIW capabilities.

The end state crafting process is likely to be an aggregation of assessments of the impact and possible future evolution of threats identified in various SIW scenarios. This process may or may not be shaped by related targeted strategy and policy decisions.

## AN INITIAL ARRAY OF POSSIBLE END STATES

Table 6.1 presents an initial array of possible alternative first-generation SIW asymptotic end states, based on the above approach. More detailed descriptions of and rationales for these end states follow:

- U.S. Supremacy in offensive and defensive SIW: The United States overwhelmingly dominates the SIW environment because it possesses

  — The world's best offensive SIW tools and techniques, which are capable of penetrating any other country's SIW defenses.

  — Highly effective SIW defenses and reconstitution and recovery capabilities that effectively reduce the vulnerability of potential SIW targets in the United States (such as key U.S. infrastructures) to strategically insignificant levels. The United States *selectively* shares these capabilities with allies.

  — Traceback capabilities that produces very high levels of confidence in U.S. abilities to identify perpetrators. No other nation has traceback capabilities good enough to identify the United States as the source if it launches SIW attacks.

**Table 6.1**

**Alternative Strategic Information Warfare (SIW) Asymptotic End States**

| Key Dimensions | A<br>U.S. Supremacy in SIW | B<br>Club of SIW Elites | C<br>Global Defense Dominance in SIW | D<br>Market-Based Diversity |
|---|---|---|---|---|
| Number of offensive SIW players | 1 | ~5–0 | 0 | $10^3$–$10^6$ |
| Degree of SIW vulnerability | U.S.: very low<br>Others: moderate-high (very high vs. US) | Among club: moderate-high<br>Club (others attack): low<br>Others (club attacks): high<br>Others vs. others: low–high | Very low | Moderate vulnerability to significant (~natural) disruption<br>*But* reconstitution/recovery fast |
| Uncertainty in perpetrator identity | U.S.: very low<br>Others: moderate-high (~100% vs. U.S.) | Club: very low<br>Others: moderate-high | Low | High |
| Tactical Warning (Is attack under way?) & attack assessment (how big & at what?) | US: very good<br>Others: less good | Club: very good<br>Others: low–moderate | Very good | Good |
| Damage assessment (size and scope of damage) | U.S.: excellent<br>Others: less good | Club: fair–excellent<br>Others: fair | Good<br>(CDC-like regime) | Very good |
| Declared/implied links to other political-military instruments | U.S.: high | Among club: moderate-high<br>Club (others attack): high | High | Low |
| Degree of international cooperation | Low | Among club: some, not a lot (NPT-like; mutual deterrence of SIW use among club) | Globally very high (within/between governments *and* infrastructures)<br>Establish BV/CW-like "SIW arms control" regime | Market-facilitated reinforced international cooperation on reconstitution/recovery |

- Club of SIW elites: Through a combination of technical capability and resource allocation, an international condominium of a handful (5–10) of highly competent SIW nations emerges. The United States is almost certain to be the most competent of the group. Within this group it would not be surprising if a smaller number (2–3) of groups of nations (allies) or strong individual nations with distinct SIW perspectives or postures emerges. Mutual deterrence of SIW use is the shared goal among club members. This handful of SIW major leaguers collaborates with each other to some degree

  — to constrain the spread of major-league SIW capability to other nations, particularly non-nations,

  — to deemphasize SIW and establish a norm of no first use of SIW (and related concepts and definitions of SIW use),

  — to set international technical standards for cyberspace that help to perpetuate the exclusivity of the club.

- Global "defense dominance" in SIW: As a result of broad global cooperation in the fielding and maintaining of very high quality SIW defenses, the vulnerability of key potential SIW targets (such as key infrastructures) in most nations is reduced to strategically insignificant levels. This end state is further bolstered by international cooperation in the global dissemination of

  — High-quality traceback capabilities (and/or a commitment to provide "whodunit?" traceback information in the event of a serious SIW attack).

  — High-quality TW/AA capabilities. This end state would also be bolstered by establishment of an SIW "arms control" regime, similar to the BW and CW arms control regimes, that establishes international information operation norms, standards, legal restrictions, and enforcement mechanisms. Like currency counterfeiting, software piracy, and other threats to world economic order, SIW becomes something responsible states do not do. SIW "rogues" are dealt with as the UN dealt with Saddam Hussein: Deny them their goals and punish them.

- Market-Based Diversity[1]: The extent of damage or disruption achievable in an SIW attack is modest, and reconstitution or recovery is fast because

  — The natural strength of diversity the globalization and standardization of cyberspace reduces overall vulnerability to SIW attack to moderate levels.

  — Global cooperation provides high-quality damage assessment tools.

  — Market-reinforced ("good neighbor") cooperation exists for reconstitution and recovery.

---

[1] The number of capable actors under this end state is very large, but so are the number of defensive solutions to apprehended vulnerabilities. Thus, many capable actors exist, but no one possesses a strategic capability in the offensive realm.

## THE FIRST IN AN EVOLVING SERIES OF FRAMEWORKS

At the outset, confidence in the sustained validity of the end states identified as worthy of consideration will be understandably low. However, over time, as IT and international relations progress and as understanding of the first-generation SIW problem improves, it will become possible to look farther ahead and gain increasing confidence in the ability to chart more realistic end states as part of the *evolving series of frameworks* process.

Keeping this perspective in mind, Figure 6.1 shows "X," which is where the SIW problem stands now with all of its uncertainties (see Chapter Five), and the alternative end states articulated above .

In considering the challenge of making near-term decisions on SIW-related issues, it is instructive to look more closely at the alternative end states included in Table 6.1 to see whether, in spite of their disparate character, they might "cluster." For example, in Figure 6.1, the C and D end states are represented as clustering because of their emphasis on cooperation and defense. Thus, in the face of the many SIW-related uncertainties, some aspects of near-term strategy and policy might be established with the goal of doing a final "reckoning" at a later time.

Later, in keeping with the concept of an evolving series of frameworks, one might find that the situation shown in Figure 6.1 ends up being more like the situation depicted in Figure 6.2—one is in a different place, and your end state options have changed.

Keeping this perspective in mind, the next chapter looks at the question of evolving series of frameworks by considering of the potential impact of a series of major perturbations in the SIW environment.

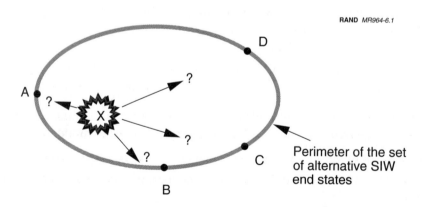

RAND *MR964-6.1*

**Figure 6.1—Where Are We (X) and Where Should We Be Going (A-D)?**

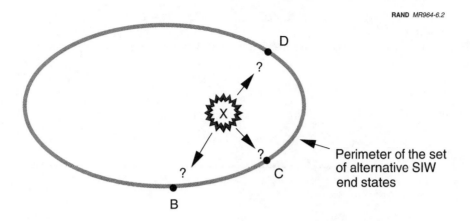

Figure 6.2—Where Might We Be in the Future?

# Chapter Seven

# AN EVOLVING SERIES OF FRAMEWORKS

## INTRODUCTION

As mentioned previously, we anticipate an evolving series of frameworks—or an evolving framework whose characteristics remain fairly stable but yield different strategy, policy, and other alternatives and decisions when revisited. It is therefore instructive to ask what large perturbations might be conceivable (but not necessarily easily predictable) in the first-generation SIW environment (such as those related to perpetrator ambiguity or encryption) that might generate the demand for the second first-generation SIW framework.

It seems likely that the demand for such a new framework will be rooted in a perceived different set of pressing strategy and policy issues and options (different from those depicted in Table 6.1). It is also possible that, with a new framework, the key dimensions of the first-generation SIW problem might have changed.

## MAJOR POTENTIAL PERTURBATIONS

During the Cold War, the strategic framework shifted several times (see Appendix A) in response to threats, political conditions, technological opportunities, military capabilities and forces, nuclear weapon experience, and intellectual sophistication. It therefore seems that one should anticipate shifts in SIW emphasis and approach, or even the emergence of entirely new SIW frameworks. Therefore, a viable initial framework needs to have features that can accommodate the volatile and uncertain forces of change. In comparison to the Cold War experience, what might these forces be and how might they be accommodated in the initial SIW framework?

Here are some possible scenarios for large perturbations in the initial first-generation SIW environment that might generate the demand for a second first-generation SIW Framework:

1. Effective use of SIW against the United States. A regional strategic crisis occurs, and an adversary actually *uses* SIW against the United States to great strategic advantage. The United States cannot resolve the crisis in a manner consistent with its public policy and a priori demands because the public response to the SIW attacks is a combination of shock and awe. The United States is seen to be vulnerable (even if the immediate loss is relatively trivial), and the crisis reaction is strategic "withdrawal" to limit losses and regroup. The United States is embar-

rassed. So much for U.S. SIW superiority, so much for deterrence of SIW attack. A new framework is needed because the last one failed (that is, the SIW decisions the United States made did not meet its national security needs).

2. Widespread strong encryption.[1]  Infrastructures protected through strong encryption and system protection techniques are likely to be significantly less vulnerable than unprotected systems, making a sustained and structured strategic IW attack on key infrastructures more difficult. Hackers and others wishing to surreptitiously enter networks, however, can also exploit limited-strength encryption technologies. The continuing feasibility of "brute force" attacks aimed at breaking strong encryption protection is intimated by the rapidity of changes in the levels of computing power deployed in society, and by progress in "parallelized" attacks—which use networked computers to break strong encryption systems—on established cryptographic products. The deployment of strong encryption *defenses* may engender an "arms race" with those intent on breaking into protected systems. In turn, hackers and others may wish to conceal their identities in response to progress in reducing perpetrator ambiguity. International collaboration on agreed-upon levels of strong encryption can be envisioned, as can discussions on coordinated investigations of suspected network intrusions. Generally speaking, however, strong defenses may make arguments for international collaboration less compelling, because options for self-reliance may prove attractive—at least to government policymakers. In such a situation, pursuit of end state A, U.S. supremacy in SIW, may prove to be the most attractive option.

3. Compromise of U.S. offensive SIW programs. A disgruntled insider, spy, or enterprising reporter reveals in broad terms the extent of U.S. offensive SIW programs. The leaks might also reveal key vulnerabilities of U.S. friends and allies, and the extent to which the United States has been "fooling around" in their systems. Worldwide reaction is very negative, and the United States is embarrassed. Offensive SIW programs are terminated, and defensive SIW gets a boost in the United States and elsewhere.

4. Russian Mafia use of SIW to sack the Russian government. Over a period of years, Russian-based transnational criminal organizations (TCOs) rob the Russian government to the point where the Russian economy collapses, and Russians suffer terribly. The fragility of modern societies to information attack and exploitation is exposed. The world resolves to help rebuild the Russian economy while pledging that the only morally acceptable use for SIW is as a deterrent. However, background debates about the usability of SIW persist. As always, the offense seems to be a step ahead of the defense, but not decisively. Practical considerations therefore continue to favor SIW as a deterrent. Those who cannot accept punitive deterrence concepts that imply mutual vulnerability and who would rather keep things under more proactive positive control continue to search for both offensive and defensive alternatives.

---

[1]The availability of strong encryption is likely to be limited by governments. Thus, the relative, not absolute, strength of encryption deployed in commercial software is the issue for the purposes of analysis.

5. Collapse of electronic commerce.  In the future, a collapse of electronic commerce could involve the collapse of both traditional bank-to-bank electronic funds transfer systems (or a significant compromise of them), or a successful penetration and disruption of consumer-level cyberpayment (electronic stored value and Internet-based value transfer) systems.  Because banking systems have the best cyberspace defenses, such an assault would likely cause governments to question the wisdom of an open systems approach to critical infrastructure operation, at least in the payment systems area.  If an electronic commerce collapse were restricted in scope to a single country or a small group of countries, coordinated action by the international community (perhaps led by the United States) might lead to a new international currency security (and perhaps monetary) system.  Because of the widespread use of electronic payment products, such a collapse would affect end state D most directly, with a shift away from market-oriented (non–state managed) solutions toward a more state-centered solution to perceived vulnerabilities.  End state B, the club of SIW elites, may be the likely result of such an occurrence.

6. Big win by United States because of overt use of SIW.  The United States uses SIW to promptly and effectively win a war or to prevail in a major crisis (similar to U.S. use of nuclear weapons against Japan to end World War II).  As a result, the rest of the world more squarely faces the SIW problem.  Allies more aggressively approach the United States for defensive SIW assistance.  Adversaries, regional and global, are stimulated to both deploy offensive SIW capability and explore the level of vulnerability of key U.S. infrastructures.

7. Big win by United States because of covert use of SIW.  The United States uses SIW to promptly and effectively win a war or prevail in a major crisis in such a manner that no one else knows that SIW was the decisive factor (in a situation similar to the "cracking" of the German and Japanese codes in World War II).  In the wake of such a success, there would be a tendency to make larger investments in SIW and to view end state alternatives such as U.S. supremacy more favorably.

8. Sino-American Cold War.  In 10 to 15 years, relations between the United States and China deteriorate sharply.  The main source of conflict is a dangerous Taiwan political-military crisis that tends to militarize the Sino-American rivalry in East Asia.  Furthermore, conflict arises because of China's attempts to gain geoeconomic hegemony in Southeast Asia and its very strong resistance to enhanced U.S.-Japan security ties.  China has become increasingly competitive in both the defensive and offensive SIW domains through the development and acquisition of increasingly sophisticated dual-purpose IT (including societal control capabilities).  China openly declares that any attempt by the United States to maintain hegemony in any sphere of contemporary warfare is intolerable, and demands that the United States agree to accept either a negotiated club of SIW elites (B) or curtail its investment in offensive and defensive SIW capabilities, thereby leading to end state C, disruption mitigation through diversity.

The above eight scenarios may or may not seem very likely.  Nevertheless, each is, in principle, possible and could have dramatic repercussions.

In addition to affecting the relative attractiveness of various end states, major perturbations might also affect other steps in the framework decision process (see Figure 2.1). The perturbations affect the inaugural first-generation framework, as depicted in the steps of Figure 2.1 as follows:

- Step 1 (key dimensions): For each of the above-mentioned major perturbations, no strong arguments were found for changing the key dimensions of Table 3.1.

- Step 2 (strategy and policy issues): Almost all of the sample set of issues discussed in Chapter Four would continue to remain relevant for the above-mentioned major perturbations, though in some cases there would clearly be a shift in emphasis.

- Step 3 (existing state of SIW): Some of the above perturbations would affect at least the perception of where one is starting from in the SIW realm as shown in Figure 6.2.

- Step 4 (alternative end states): As noted above, several of the perturbations would, at least change the priority or level of interest in the four end states depicted in Table 6.1. Whether in some circumstances certain alternatives might be eliminated altogether is a matter of debate. For example, if both the first and third of the above-mentioned major perturbations took place, end state A might be expected to be of much lower interest; alternatively, such experiences might produce a new strategy commitment to achieve U.S. SIW supremacy and a marked increase in resources allocated to this goal.

- Step 5 (alternative action plans): Most of the major perturbations discussed would catalyze new action plans both in the United States and elsewhere, some of which were speculated on in describing the perturbations.

## THE FRAMEWORK AS A MEANS OF SHAPING THE FUTURE

A number of perspectives need to be kept in mind regarding the use of the SIW framework presented in this chapter. First, the framework should facilitate better shaping of the SIW environment so that the more dreadful of our imagined scenarios become much less likely. To do this, the imagination must be allowed to "run free," to create challenging SIW scenarios and then attempt to categorize these scenarios by common characteristics and dimensions that can be addressed in consistent policy initiatives.

Second, one can build in branch points and "sign posts" to allow incremental and prompt adaptation to environmental changes without discarding the whole framework and "starting from scratch." For example, although the initial framework might build on assumed U.S. SIW superiority, the United States would need to identify and monitor specific indicators that would tend to refute this underlying assumption, and think through response options if troubling indications are detected.

Finally, one can use the uncertainties in the initial framework to help evaluate and prioritize possible initiatives. For example, if certain initiatives tend to be important regardless of uncertainties, as one understands them, these might be a good place to start. As resources permit, the United States can expand its efforts within the framework, keeping the investments consistent with likely payoffs in the uncertain future.

# Chapter Eight

# ALTERNATIVE ACTION PLANS

As discussed previously, the fashioning of an SIW action plan involves (1) gaining the best possible reading on where one is, and then, within the limits of uncertainties, (2) charting a course or pathway for a near-or long-term goal or end state (however roughly defined) through the choices made on a carefully chosen set of policy, strategy, and related issues.

With this perspective in mind and as a final presentation of potential outcomes, Table 8.1 arrays those key issues identified in Chapter Four as potentially ready for near-term decisionmaking in terms of which options for these issues favor which end states. A comprehensive examination of such alternatives—and an initial set of choices—appears to be a priority task for U.S. government and industry in forging an initial action plan in this problem area.

Some additional perspectives on the individual issues and the array of presented end states follows:

1. Locus of responsibility and authority. This issue, arrayed against the various end states, must be faced immediately. Can the SIW problem be left to the marketplace to solve, or is some fundamental government intervention required? If there is a serious SIW-related national security threat—and it appears at this time that there could be—some level of government involvement appears to be imperative. RAND's exercises to date and the report of the PCCIP argue strongly for some kind of joint responsibility. Whether on the government's side this should have a national security orientation or a law-enforcement orientation remains debatable and is likely to be very much affected by preferences for the end state (B versus C).

2. Tactical warning and alert structure. This issue is also strongly affected by the degree of the threat from nations that could mount well-coordinated and well-structured SIW attacks. As indicated in Table 8.1, a number of different warning and response models might be considered, again depending in part on the preferred asymptomatic end state. The PCCIP report, the Defense Science Board (DSB) Information Warfare report, and the results from the RAND exercises also support creating such a warning as a priority matter if there is to be any serious effort to combat the SIW problem. Creating such a system, which cannot be done without substantial government-industry cooperation, may be a critical means of fostering the government-industry cooperation that is needed across the board on this subject.

**Table 8.1**

**Alternative Action Plans**

| Key Strategy and Policy Issues | Competition | Mixed (Competition and Cooperation) | Cooperation | |
|---|---|---|---|---|
| | A | B | C | D |
| | U.S. Supremacy in SIW | Club of SIW Elites | Global Defense Dominance in SIW | Market-Based Diversity |
| Locus of responsibility/authority | Federal government leads; national security focus | Federal government leads; national security focus | Federal government leads; law enforcement focus | Industry leads |
| | Joint leadership | Joint leadership | Joint leadership | |
| Tactical warning and alert structure | Government-led NICON model | Government-led NICON model | CDC model | Industry-led model |
| | Counterterrorism model | Counterterrorism model | Industry-led model | |
| | | CDC model | | |
| Declaratory policy (links with other military instruments) | Strong retaliation threat (SIW retaliation emphasis) | Moderate retaliation threat vs. non-club actors | No retaliation threat | Moderate retaliation threat (emphasis on economic instruments) |
| | Reassurance on invulnerability of key U.S. infrastructure | Some reassurance on invulnerability of club NIIs | Reassurance on resilience of GII | |
| International information sharing and cooperation | SIW programs compartmentalized | High degree of cooperation within club (G-7/FATF model) | High degree of cooperation | High degree of voluntary cooperation |
| | | | Institutional links through NATO, FATF, etc. | |
| Vulnerability assessments | Government-led (NICON organizational model) | Government-led (G-7/FATF model) | Public/private U.S. (WHO model) | Public/private U.S. (CDC model) |
| R&D investment strategy priorities | National security protection and coordinated alliance action | Coordinated international action and proscriptions on offensive SIW research | Coordinated R&D, with offensive R&D proscriptions and a private sector focus | Private sector focus, with proscriptions on offensive SIW R&D |

3. Declaratory policy. Declaratory policy constitutes a major problem because of the profound uncertainty in perpetrator traceback and identification techniques. This issue will be affected by the outcome of the national and global encryption debate. It therefore seems unlikely that any clear and temporally stable declaratory policy statement will be achievable in the near future. The implicit (possibly made explicit) posture on this issue appears to be ambiguity about the response to SIW attack. Explicitly, this posture of ambiguity (which could also be the chosen posture even if traceback techniques improved dramatically) would likely be couched in terms that threatened retaliation with any or all the available instruments of strategic leverage, from strictly military to economic instruments.

4. International information sharing and cooperation. This issue emphasizes the challenging problem of international information sharing on defenses against SIW attack. To adequately address this issue, careful consideration must be given to the desired SIW end state. If either the U.S. Supremacy in SIW (A) or the club of SIW elites (B) end state is seen as both desirable and achievable, sharing of such information will be highly restricted. In contrast, the other two end states assume, if not demand, a high level of global cooperation.

5. Vulnerability assessments. Some means of obtaining an assessment of comprehensive infrastructure vulnerability is imperative. But as shown in Table 8.1, who should do it and how such information is distributed and integrated is a function of the preferred end state.

6. R&D investment strategy priorities. This issue is driven largely by whether the United States is interested in maintaining a strong SIW capability (end state A) and whether the United States (and/or possibly a group of like-minded nations) sees perpetrator identification possible and therefore makes direct retaliation against SIW attackers a plausible SIW posture (end state B). However, if cooperation and a focus on defense and reconstitution is viewed as preferable (C or D), there could be a proscription on offensive SIW and global coordination of R&D similar to a public health model.

# EXEMPLARY FIRST- AND SECOND-GENERATION SIW ESCALATION SCENARIOS

## INTRODUCTION

The purpose of this appendix is to stimulate thinking about escalation in SIW-related crises and to proffer exemplary first- and second-generation SIW escalation scenarios. The issue of escalation—the potential increase in intensity and severity of a political crisis—arises with the use of SIW, just as it does with other forms of warfare. SIW tools and techniques therefore add both quantitatively and qualitatively to conflict dynamics. On the one hand, escalation in SIW-related crises might be made more controllable, as decision makers gain the ability to signal their resolve and intentions through the use of SIW instruments. On the other hand, SIW techniques may have the potential to destabilize the management of international conflict, as third parties gain a greater ability to intervene in conflicts between nations, perhaps worsening relations and diminishing the prospects for a settlement of differences without actual violence.

The discussion below describes and assesses two exemplary scenarios:

- First-generation SIW "China/Taiwan 2010" scenario. This regional strategic crisis scenario examines the complexity of weighing SIW use against other more traditional military instruments.

- Second-generation SIW "Russia 2000–2010" scenario. This scenario describes a long-term Russian SIW campaign for which the objective is to use SIW instruments to gain dominance over the global petroleum and natural gas supply system and market.

These and similar scenarios emphasize that, to understand the potential role of SIW in a crisis, it is crucial to recognize that SIW instruments can be used creatively in combination with more traditional tools of international relations.

## ESCALATION CHART METHODOLOGY

When considering the prospect of SIW use by nations (and other entities) in pursuit of political objectives, it is instructive to chart the concept of escalation over time in some graphical manner, as provided in Figure A.1.

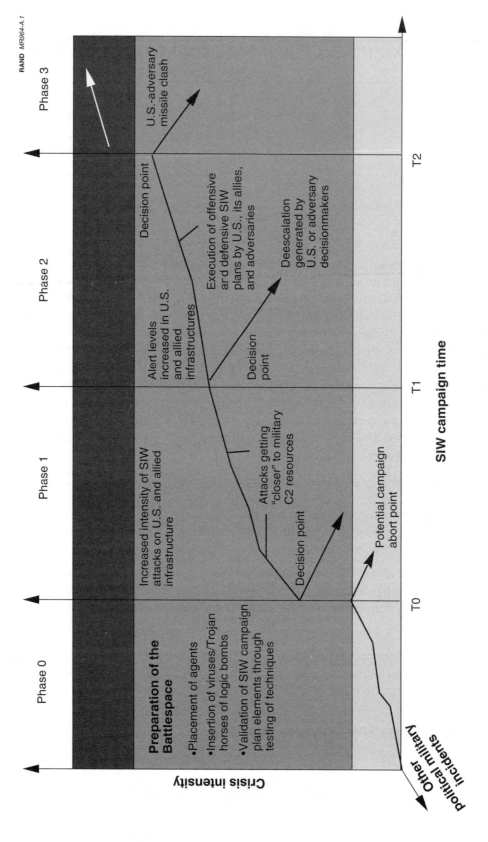

Figure A.1—Overview SIW Escalation Chart

The vertical axis of Figure A.1 labeled "Crisis Intensity" indicates the political seriousness of an SIW crisis (for example, as perceived by elite decisionmakers). This subjective notion of crisis intensity is portrayed in three distinct step—green, yellow, and red—as indicator of increasing infrastructure alert status. As one moves up the vertical axis from a Crisis Intensity state of green to yellow to red, a number of governmental and private sector alert activities would presumably occur:

- Additional alert measures would be taken, involving the deployment of law-enforcement, IC, and other governmental and private sector infrastructure security resources.

- The political stakes at risk in further actions during the crisis would increase, as concerns about deterrence of offensive actions and escalation reinforce the underlying material stakes at risk.

- Top-level decisionmakers would become more and more involved as the crisis increasingly displaces other political issues from the top of the agenda.

The horizontal axis labeled "SIW Campaign Time" reflects the time period over which the SIW campaign evolves. As shown, this axis is discontinuous. The actual pace of events will reflect the success or failure of SIW techniques in achieving their initiator's objectives, and the difference between predicted and actual responses by the nation defending against SIW assault.

The achievement of intermediate goals within a campaign is key to determining SIW campaign duration and character. At each decision point in Figure A.1 (that is, T0, T1, and T2) the actors involved would evaluate the situation in terms of both the stakes at risk and their respective objectives (both defensive and offensive). A deescalation move by any actor can be reinforced by adversary behaviors or opposed by them, with a consequent significance for the overall crisis.

A "Z" axis, presented as a third dimension orthogonal to the Crisis Intensity and SIW Campaign Time axes, represents the use of more traditional military instruments; the relationship between that domain and the SIW arena is one of the critical issues explored within this project. In particular, the question of how escalation or deescalation in hypothetical SIW conflicts affects traditional political-military activities and the use of other military instruments is a critical question.

## FIRST-GENERATION SIW ESCALATION SCENARIO: CHINA/TAIWAN 2010

### Overview

The political and military leadership in Beijing became "fascinated" with the concepts of SIW after examples of successful SIW use in other conflicts. Of particular importance was the use of SIW tools and techniques in the Persian Gulf crisis of 2006 and the 2007 South Asian conflicts. In the Persian Gulf, Iran, while failing to meet all of its strategic objectives, had succeeded in impeding U.S. diplomatic freedom of action in the projection of power to the region. This example convinced members of the Chinese leadership that they might be able to achieve a similar objective in East Asia.

## The South China Sea Crisis of 2008

In late 2005 China landed marine forces on the Spratly Islands. Simultaneously, it launched a carefully calibrated SIW campaign against Malaysia, the Philippines, and the United States designed to (1) demonstrate Malaysian and Philippine vulnerability and thereby reduce alliance cohesion and the credibility of United States pledges to protect allies, and (2) signal strongly (by engaging U.S. infrastructures in a way that activates warning systems, but does not activate a strategic overreaction) that China could affect U.S. territory if the United States chose to intervene in the Spratly Islands dispute.

After an air and naval clash with Chinese forces in the South China Sea, the Malaysian, Philippine, and Japanese governments all protested what they claimed was a Chinese attempt at a *fait accompli* on the Spratly Islands issue. They asserted their nonrecognition of any forcible redrawing of territorial boundaries in Asia.

The United States remained largely passive throughout the crisis, which ended with regional and U.S. acceptance of Chinese authority over the Spratly Islands. U.S. and East Asia media spoke of a "strategic victory" by China that was "as significant as the Japanese defeat of the Russian fleet in 1905."

As the result of the Spratly Islands crisis, three major strains of Japanese thinking on relations with the United States crystallized: (1) those who believed that China could not be contained without an arms race or without a continuation of Japan's close political and military relationships with the United States, (2) those who felt that Japan needed to become a "normal state" taking unilateral responsibility for its defense and security relations with its neighbors and with the rest of the world, and (3) those who saw the U.S.-Japan relationship as fading, thereby requiring that Japan come to terms with the rise of China.

After considerable domestic debate, the U.S. administration concluded that it could tolerate neither Chinese hegemony over East Asia, nor the prospect of a full-blown strategic competition between China and Japan. Therefore, the administration declared that it was prepared to use force to resist any effort by mainland China to forcibly reincorporate Taiwan.

## 2010: Phase 1 Situation Report

A widespread belief emerged among Chinese elites that the United States and Japan were actively supporting political and economic efforts by Taiwan to gain autonomy, if not outright independence. After a period of rising tensions, China announced that it had evidence that Taiwan was on the verge of acquiring an operational nuclear capability and backed up this charge by mobilizing its major air and sea units, clearly intended to signal Beijing's willingness at this point to isolate Taiwan physically through an air and naval blockade.

In response to this Chinese show of force, the United States reinforced its naval and air posture in the Western Pacific. China responded by firing several missile barrages near the major ports of Taiwan and threatening a blockcade to compel the Taipei government to halt its nuclear weapon program and acquiesce to major political and economic concessions that would thwart independence.

Further adding to the crisis was a dramatic increase in SIW-related incidents within the United States and Japan. Even though both countries had enhanced tactical warning and attack assessment systems, the source of the attacks were not readily identified.

When it became evident that the United States and Japan were not intimidated by either their military maneuvers or early SIW attacks aimed at homeland targets, the Chinese leadership reached a crossroads (see point "T0" in Figure A.2):

- One faction advocated a "cooling-off period" and argued that a further escalation of the crisis would only solidify the American-Japanese political military alliance into an anti-Chinese coalition.

- The second faction argued for escalation in the belief that neither the United States nor Japan were prepared to risk a major military clash in East Asia over Taiwan.

## 2010: Phase 2 Situation Report

The U.S. Administration faced a number of strategic choices in the worsening crisis (see point "T1" in Figure A.2):

- Take only NSI defensive measures and continue the military buildup in the Western Pacific.

- Consider an SIW "counterattack" against Chinese NSI targets

- Take only NSI defensive measures, and attempt to begin a negotiated end of the crisis with Beijing.

After considerable debate the Chinese chose to escalate the crisis. An air and naval blockade was declared around Taiwan. This "no fly/sail zone" extended some 100 nautical miles north, south, and east of the island. Beijing announced that the blockade would continue until Taiwan accepted a negotiated solution to Taiwan's absorption by China.

The Taiwan government was defiant and vowed to "break Beijing's monstrous act of international piracy." In the United States, the President announced that "all measures would be taken to break the Chinese blockade of Taiwan" as a "clear threat to international peace," demanding that China accept an Indonesian proposal for an East Asian regional summit to resolve this crisis. Additional air and naval reinforcements were sent to the Western Pacific, including the deployment of a second wing of heavy bombers to Guam.

In Japan, the government asked its police authorities to increase their surveillance of anti-U.S. groups and began informal discussions with the major electric power and

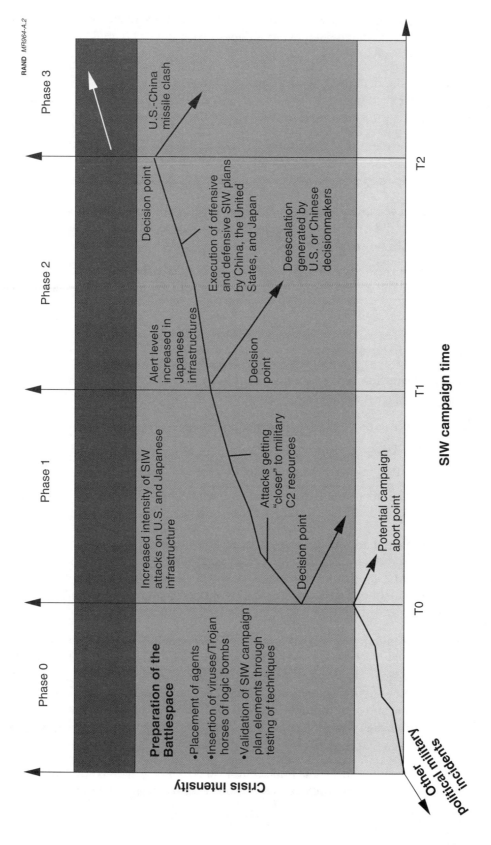

Figure A.2—China-Taiwan 2010 Scenario Escalation Chart

rail operators on cyberspace security matters. China chose to escalate its SIW campaign, using conventional military activities designed to deter U.S. intervention. Test flights of a new intercontinental ballistic missile (ICBM) were conducted across the Pacific Ocean, with their warheads reentering the atmosphere 800 miles west of Hawaii.

Taiwan's first attempt to run the Chinese blockade ended in failure when two large container ships escorted by several Taiwan frigates and maritime patrol aircraft were attacked by missile-equipped Chinese long-range aircraft and submarines. Both container ships and three frigates were sunk. Both air forces suffered substantial losses.

Hours later, a U.S. SSN was attacked by a Chinese submarine, which was sunk in the ensuing engagement. A new Taiwan attempt to break out of the Chinese blockade with air cover from two U.S. carrier battle groups led to a major clash with Chinese land-based aircraft and submarines. After the Chinese suffered heavy losses from U.S. naval air intervention, the U.S. carriers were subjected to a Chinese ballistic missile and submarine attack. One carrier suffered modest damage from two near-misses, but three escorts and an arsenal ship suffered serious damage, prompting the task force commander to withdraw the U.S. fleet east of Taiwan.

In Japan, security agencies responded to the Chinese SIW strategy to increase the influence of those who wished to alter the historical relationship with the United States by launching an assault on a list of known anti-U.S. and pro-China groups, obtaining authority to trace network traffic emerging from sources known to provide services to suspected groups. At the same time, the Japanese government ordered the rail, financial, and electric power grids to raise their level of alert, with critical systems to be taken off-line for the duration of the emergency. In addition, electronic trading of securities was curtailed for the remainder of the crisis. Japan also initiated consultations with the United States regarding possible coordinated action against China.

In Washington, in the wake of the U.S.-Chinese air and naval clashes, the President examined several strategic options (see point "T2" in Figure A.2):

- Continue military operations to break the Chinese blockade including the use of SIW techniques to profoundly degrade the Chinese military command and control system.

- Change the nature of the conflict by using a major SIW campaign against the Chinese national strategic infrastructure—with or without restrictions on attacks on the Chinese financial system.

- Call for a cease-fire and attempt to move the conflict rapidly to a negotiating track.

In Japan, several options were considered by the government:

- Call for an immediate cease-fire, and move the conflict to a negotiating track.

- Consider launching a major SIW attack against the Chinese infrastructure.

- Take only defensive measures to protect Japanese infrastructures from further SIW attacks.

- Ask the U.S. not to use its air and naval facilities in Japan in support of further military operations against China.

In China, disagreements within the Chinese leadership had become acute. Opponents of the original decision for escalation argued that the apparently imminent regional air-naval war with the United States would have disastrous results. The militants in the leadership were adamant that backing down would result in a humiliation as grave as China's defeat by Japan in 1938, with negative implications for China's emergent status as a great power. The militants continued to advocate an escalation of the SIW campaign, focused on breaking the will of the Japanese leadership as the path to making the American position in East Asia politically and strategically untenable.

## SECOND-GENERATION SIW SCENARIO: RUSSIA 2000–2010

### The Global Economy and Oil Market

Although the major industrial democracies in North America, the European Union, and East Asia have taken significant measures to increase energy efficiency, the worldwide demand for petroleum products continues to steadily increase. Principally responsible for the increase are the booming economies of East Asia, particularly China. Although there have been major efforts with noteworthy success in oil and natural gas exploration to diversify sources of supply, known global reserves remain heavily concentrated in the Persian Gulf, Russia, and Central Asia.

In the midst of this situation, an authoritarian and nationalist Russian government elected in 2001 concluded that Russia had a historic opportunity and geostrategic necessity (as a means of shortening the period of post–Cold War economic recovery) to gain great influence, if not dominance, over the global petroleum and natural gas supply system and market. Russia based this decision on the heavy dependency of the global economy on the orderly flow of petroleum products at non-inflationary prices from selected global locations. The opportunity envisioned was the manipulation of international raw materials and derivatives markets via SIW techniques in a way that would ensure for Russia a larger share of revenues from raw materials trade throughout the first decade of the 21st century.

### Moscow in 2001

Preparation of the raw materials SIW campaign before the onset of the actual program to manipulate international commodity markets (see escalation chart in Figure A.3) involved extensive intrusion into protected communications and computing systems. As Figure A.3 indicates, the exact initiation point of the "preparation of the battlespace" phase is unclear. Detection of systematic alterations in the behavior of systems that are assumed to be secure is especially difficult. Russia pursues a com-

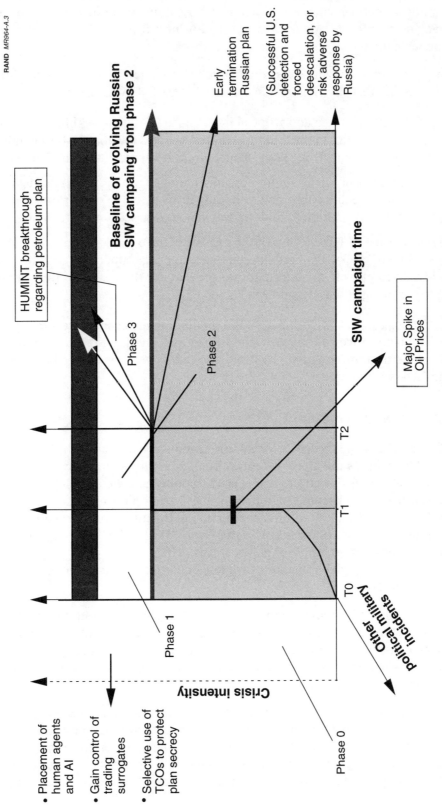

Figure A.3—Russia 2000–2010 SIW Scenario Escalation Chart

plex multiyear strategic campaign including manipulation of electronic (space) futures and commodities markets, diplomatic bargaining with other oil producers, influencing of trading and pricing decisions by business firms, and the use of disinformation techniques.

The first phase of the Russian SIW campaign to manipulate electronic commodities and derivatives markets (see Figure A.3) consisted of the following measures:

- The placement of human and artificial intelligence (AI) agents within the trading systems of firms active in the trading of derivatives and raw materials in London, New York, Paris, Rome, Hong Kong, Singapore, and Tokyo (Phase 0 in Figure A.3)

- Gaining of control of surrogates for implementation of the campaign through the selective purchase and suborning of trading firms active in one or more of the major derivatives and commodities trading venues (Phases 0 and 1 in Figure A.3)

- The selective use of TCOs to ensure the security of the overall program, including selective use of blackmail, murder, economic espionage, and other activities that would help to provide the information necessary for the design of the market-shaping campaign (Phases 1 and 2 in Figure A.3).

Russia's "strategic petroleum campaign" is a predominantly cyberspace affair, seeking to manipulate prices and supply conditions over the long-term to finance a strategic modernization program.  By 2009, this campaign appeared to be partially compromised (see below).

## 2002: Phase 1 Situation Report

By 2002, Iran's enhanced prospects for gaining hegemony over Iraq in this timeframe was viewed by many of the Moscow "elites" as a potential geostrategic and geoeconomic windfall.  Iran is viewed as a useful ally in gaining control over oil production, largely because Iran has become an important geostrategic ally to Russia and a major buyer of Russian arms.  Furthermore, a reduction in the intervention capabilities of Western Europe and the United States in the Persian Gulf will help Russia regain its "feet" in the "peer competition" for global stakes.  Thus, anything that impedes the West from controlling global markets promotes long-term Russian economic and geostrategic interests.

A faction within the Moscow regime therefore strongly supported an aggressive diplomatic posture in support of Iran during Tehran's quest to gain hegemony over Iraq.  Proponents of the "strategic petroleum plan" were more cautious and feared any active Russian role on behalf of Iran might compromise Russia's long-term strategy to manipulate the petroleum market.

During the Iraqi civil war crisis in late 2002, the authors of the strategic petroleum plan were very nervous that their long-term strategy would be compromised by actions taken by Tehran.  However, Iran was successful in intervening in the civil war in Iraq and installing a friendly government in Baghdad, while deterring any decisive counter-response by either Turkey or the United States.  Through a global disinformation campaign, Moscow was able to dampen speculation in the media about whether the Russian government had acquired significant "windfall profits" during the crisis by design.

## 2005: Situation Report

Much to the consternation of the Russian practitioners of the strategic petroleum plan, Iran renewed its regional geostrategic offensive after consolidating its position in Iraq. The target of Iran's effort was to overthrow the Saudi Kingdom, still a linchpin of the U.S.'s continued effort to contain the Persian state geostrategically. Although the Moscow regime continued to maintain a close political, military, and economic relationship with Tehran, there was broad concern that the "Persians were overreaching" in their bid to destabilize the Saudi Kingdom while expelling the U.S. political-military presence in the Persian Gulf region.

Throughout the crisis of 2005, Moscow took a relatively low profile while giving Tehran strong diplomatic support in the U.N. Security Council debates. Only an "11th-hour" diplomatic intervention by the Russian Federation and China saved Iran from a massive air and missile attack during the height of the crisis.

Although the Saudi Kingdom survived the attempted overthrow, global traders in petroleum were profoundly shaken by the prospect of a major disruption in petroleum supply. Unlike the price spikes that followed the 1991 Persian Gulf War, and the Iraqi crisis of 2002, petroleum prices did not return to their preconflict levels, but instead remained elevated by more than 30 percent during the late 2000s. As Figure A.3 indicates, the end of this phase (decision point T1) saw a significant increase in world oil prices—ostensibly the result of the still covert strategic petroleum plan.

## 2009: Phase 2 Situation Report

The Russian SIW campaign from 2005 to 2009 continued to consist of tactical trading and TCO-related intimidation moves calculated to protect the integrity of the implementation of the strategic petroleum plan, and of tactical shifts in the trading models and trading strategy bias to respond to changes in market conditions and the entry of new traders. Throughout this period, the global petroleum price fluctuated around a trading range that remained 30 percent higher than the previous five-year average.

In 2009, contrary to the instructions of the Russian regime, several TCOs in Russia chose to engage in major cyberspace (SIW) attacks against financial system targets that resided within Russia, Poland, and Germany. Evidence that several major Polish and German banks were vulnerable to very sophisticated penetration and attack prompted a major global run on the mark and the zloty. The resulting financial panic led to two days of instability in the value of the euro, the European Monetary Unity currency. Regulating bodies in Germany, London, and Paris detected evidence that Russia was exploiting the market instability that followed in the global commodity markets.

To further complicate this situation, a HUMINT source confirmed the suspicions of many central bankers and market regulators in Europe, North America, and the Far East that Moscow had authored a massive attempt to manipulate the petroleum and other commodity markets. This is the only point at which the infrastructure-specific SIW tactical warning and alert systems succeed in detecting the departure from normal network functioning—as indicated by the transition at decision point T2.

### 2010:  Phase 2 Situation Report

Although Moscow faced numerous threats of economic and financial retaliations for its act of "criminal mercantilism," the deteriorating situation in the Far East had become a major distraction for the United States. Although the Chinese leadership expressed some private annoyance at Moscow's successful effort to enrich itself at China's expense, Beijing remained appeased in part by the continued wide-scale transfer of advanced military technology from a wide array of Russian firms. Again, as Figure 2.3 illustrates (end of phase 2), the strategic petroleum plan ends only after a serendipitous HUMINT breakthrough. The important fact is that most (in terms of time) of the strategic petroleum plan went undetected because its dynamics did not depart from the expected behavior of the information-dependent infrastructures within which the plan operated.

### CONCLUSIONS

Both of the above scenarios articulate a strategic rationale for launching a campaign that uses SIW tools and techniques to achieve a clearly visible set of objectives. The first scenario involves conflict over territory and status within a regional and global context. In this situation, SIW instruments are seen in combination with more traditional political and military tools as potential levers for influencing the outcome of a crisis. The potential use of SIW instruments to signal resolve to a potential adversary, as well as their potential use as a weapon of war, illustrates that SIW tools and techniques may have utility in disputes among nations. However, the potential ambiguity of signaling with SIW instruments shows that these tools do not end the "fog of war," but in fact may exacerbate it in a rapidly evolving crisis.

The second-generation SIW escalation scenario involves a more long-term strategic campaign aimed at achieving economic objectives. This scenario represents the type of maneuver that nations (and potentially, non-nation actors) might choose to pursue in an environment in which the economy and society of the globe has become extremely dependent on electronic funds and services flows through cyberspace. Under such circumstances, it may be anticipated that criminals might seek to use cyberspace to conceal illicit activities, such as money laundering and fraud. More narrowly, the possibility exists that nations might attempt to use criminal organizations to achieve objectives through cyberspace that could not be pursued in a more overt manner. The Russian strategic raw materials campaign outlined here is just one example of what may be a broad family of threats to infrastructure vulnerabilities with strategic consequences.

# HOW TO USE THIS TOOL

## INTRODUCTION

The test of the SIW framework as a strategic analytical tool will be its utility as a pro-cessing mechanism to systematically develop the relationship between major strat-egy and policy initiatives and short- and medium-term strategic objectives (end states). Furthermore, the framework is specifically designed to accommodate to major changes in the global security environment that either open up or preclude, on cost and desirability grounds, future end states. This analytical tool will help the decisionmaker to understand both the implicit and explicit relationships between major policy issues and strategic objectives in a highly dynamic geostrategic, geo-economic, and technological environment.

The use of potentially favored end-states in the SIW framework helps make clear the underlying assumptions of government policies that are significant for SIW. A deci-sion to pursue a particular end-state carries with it explicit requirements—and nec-essary tradeoffs—among a number of different objectives. These tradeoffs are fur-ther revealed through the use of this analytical tool to assess the key strategy and policy issues that underlie the SIW problem.

In this chapter, we will use two examples to show how the SIW framework can be used to examine the relationship between major policy initiatives in the SIW domain and strategic end states. The examples are (1) tactical warning and alert structure, and, (2) R&D strategy priorities. Each of these represents a nexus of key strategy and policy issues within the SIW analytical framework. Both sets of issues have consider-able significance for any near-term action plan that proposes any initiatives by the federal government in cooperation with state and local governments and the owners and operators of the national strategic infrastructures.

## TACTICAL WARNING AND ALERT STRUCTURES IN ALTERNATIVE END STATES

Protecting national security has traditionally been the principal job of the federal government. In the current potential SIW environment, the determination of who within government should lead in protecting the United States from foreign (and, potentially, domestic) threats is the subject of considerable controversy. Alternative conceptual models of tactical warning and alert structures highlight different approaches to the reconciliation of the government's legitimate role in protecting public safety and national defense and of the private sector's concerns about increasing economic efficiency through the rapid exploitation of IT. Each of the alternative end states summarized in Table 6.1 represents a different balance among the competing public and private objectives.

## MODELS OF TACTICAL WARNING AND ALERT STRUCTURE

Extensive analysis of the scenarios of the SIW problem reveals at least four critical models of tactical warning and alert structure: (1) the national infrastructure condition (NICON) model, (2) the counterterrorism model, (3) the CDC model, and (4) an industry-led model. Each of these models represents a different balance of public and private sector activity. The organizational structure of each of these models also differs as one moves from a more federal government-centered model (the NICON) toward a more decentralized industry-oriented framework (the industry-led model).

### The NICON Model

The NICON model has parallels with the strategic nuclear DEFCON model. It defines a series of information and institutional links that comprise a warning and alert system for national information-dependent infrastructures. The model articulates a distinct national security focus for infrastructure protection, with the DoD having a predominant role in infrastructure warning and alert planning. Defined in terms of the key dimensions, the NICON system is a joint governmental and private sector institution, with the government leading efforts in securing national strategic infrastructures from disruption and/or destruction.

The principal actors in the model are sector-specific infrastructure emergency response centers (ERCs). These centers are composed of infrastructure operators (the private sector owners) and federal oversight authorities. The centers maintain routine surveillance over their infrastructures, noting any departures from normal functioning and forwarding the information to a fusion center charged with analysis and dissemination of data on infrastructure operation and behavior. The federal information fusion and analysis center has the responsibility for pointing out any patterns of infrastructure disruption that may affect critical systems or functions (such as, electric power, the water supply, or the air traffic control system). In the NICON model, the sector alert and warning function is performed in a collaborative public-private institution.

The potential therefore exists for a subtle conflict of incentives within an institution charged with promoting the economic health of an infrastructure (the infrastructure operators) and one responsible for ensuring the effectiveness of infrastructure protection efforts. Risk orientations and definitions of "acceptable" risks and vulnerabilities can both be affected by decisions driven predominantly by economic factors. No conflict is necessary between public and private sector interests in infrastructure protection, nor is there a predictable commonality identity of interest. The NICON model manages this issue by allowing the government to mandate changes in infrastructure alert status on the basis of information received and analyzed at a federal information fusion and analysis center. Government administrative control ensures that infrastructure alert response is based on a coordinated effort to minimize uncertainty about the authorship and motivation of disruptive and/or destructive events in the different infrastructures. Depending on the level of alert declared by the individual infrastructure ERCS and/or the federal fusion center, federal, state, local, and private defense and reconstitution responses would be initiated in a structured and preprogrammed manner. Thus, warning and alert responses are tightly coupled, allowing for rapid and (presumably) effective government and private action to mitigate the damage inflicted by an infrastructure attack.

## The Counterterrorism Model

The Counterterrorism model emphasizes an law-enforcement approach to infrastructure protection warning and alert. In contrast to the NICON model, this approach highlights leading roles for federal and state (and, potentially, local) law-enforcement agencies with strong connections to the IC. Unlike the NICON system, the role of the DoD would be limited to providing assets only during those extreme cases in which it was called upon by the federal LEt agencies. This perspective views infrastructure protection as almost completely a problem of crime, though at a technologically sophisticated level. The private sector would provide information to oversight authorities under legal and administrative mandates. Staff involved in the infrastructure warning and alert response role would be professional law-enforcement and intelligence agency personnel possessing specialized knowledge and skills pertaining to network vulnerabilities and countermeasures. Ongoing safety and security measures would continue to be taken by the owners and operators of the national infrastructures.

Mandates for changes in infrastructure alert status (in response to a perceived attack or in apprehension of disruptive or destructive events) would be established in a "top-down" manner, with federal authorities determining the timing and quality of alert response measures. The implementation of alert response decisions would, however, be confederated, with infrastructure operators implementing the actual changes in system behavior dictated by the federal government. A model of this sort of response is provided by the Bank Secrecy Act's reporting requirements, which banks and money services businesses must comply with. Advisories announced by financial industry regulators are implemented in the business procedures of the industry, and with compliance data are sent to oversight agencies for systematic or random audit. In this model alert, warning and response systems would remain strongly coupled, but the implementation of responses would be undertaken by the private sector.

A major difference between the NICON model and the counterterrorism model is based on the assumption that the most plausible threat will remain in the domain of criminal behavior. This difference, though one of degree, is key to the issue of the timeliness of response in the case of an apprehended attack by domestic or foreign entities. How rapidly would alert mandates be implemented? Who would define acceptable implementation time periods? What sanctions would be imposed on infrastructure operators that failed to respond in time? Each of these questions arises as soon as a tactical warning and alert structure moves away from a government-centered approach.

## The CDC Model

The CDC model of tactical warning and alert structure constitutes the least familiar approach to detecting and responding to disruptions in system performance. Modeled after the CDC headquartered in Atlanta, Georgia, the government acts as a coordinator for detection of and response to infrastructure disruption. (Note: The federally funded but academically operated CERT represents another variant of this concept).

The CDC would be proactive in catalyzing public and private sector vulnerability mitigation measures. Professional staff government agencies, the academic community, and private sector infrastructure operators would implement a joint oversight initiative designed to detect any perturbations in "normal" infrastructure functionality. A group decisionmaking process with significant private sector (business and academia) participation would coordinate recommendations for alert responses that would include defensive and reconstitution options. Responsibility for reaction to these warning systems would lie primarily with the owners and operators of the national strategic infrastructures. As with the CDC model, there would likely be a formal tie with the national equivalents, especially allied advanced technology countries, as well as possible international organizations similar in concept to WHO. Government mandates would recede under this model to a supporting role, necessary only if isolated elements of critical infrastructures failed to respond in a timely manner to warning of major disruptive incidents.

## The Industry-Led Model

For this model infrastructure sectors themselves would assume the responsibility of coordinating protection activities. Government's role under this model would be primarily advisory, with some international consultative activity in areas of national security importance (that is, areas of importance to U.S. economic or political interests overseas, for example, Saudi Arabia and Kuwait). The principal actors in this model are the commercial infrastructure operators and whatever governmental oversight bodies exist (at the federal, state, and local levels). CDC- and WHO-like organizations with weaker mandates might co-exist under this model. Professional staff working for the infrastructure sectors would form the principal ongoing oversight and threat characterization system for the United States.

Industry might adopt a "self-regulation" approach to infrastructure protection, and a corporate body, perhaps an industry association, would adopt a supervisory role over particular commercial institutions. An example of this sort of oversight system is that of the National Association of Securities Dealers (NASDAQ), which conducts supervision of its member institutions under the auspices of the Securities and Exchange Commission (SEC). Because of the possibility of wide variations in the levels of corporate concentration (and pattern of ownership) across different infrastructures, it is possible that oversight might be undertaken in (1) a hierarchical manner, with industry associations imposing strong guidelines on infrastructure operation, or (2) more advisory-centered approaches, in which levels of infrastructure surveillance and protection are negotiated in a "standard-setting" manner among sector owners. At the international level, collaboration between national and multinational owners and operators of large information infrastructures might well create an international clearinghouse and regulatory organization, with features similar to Central Bank coordination operated through the Bank for International Settlements (BIS).

## ALTERNATIVE ACTION PLANS FOR TACTICAL WARNING AND ALERT STRUCTURES

What issues govern the appropriateness of a particular model for the United States? At the outset, a close relationship is posited between the overall strategy governing SIW and the framework used to secure U.S. information infrastructures from attack. Table B.1 portrays such a set of linkages by relating the role of the U.S. government in mitigating infrastructure vulnerability to the national strategy that the United States is pursuing in the SIW domain.

### U.S. Supremacy in SIW

To accomplish this end state, the United States has decided to pursue a comprehensive path of supremacy in both defensive and offensive SIW. U.S. government coordination and priorities dominate the selection of modalities for infrastructure protection, the coordination of action to mitigate known vulnerabilities, and the guidance under which countermeasures against attacks are implemented. This end state foresees two dominant types of SIW threat: those deriving from structured attacks of a national adversary (that is, a "near-peer" or "theater-peer "competitor), and assaults with possibly strategic consequence by other actors (such as terrorists or organized crime organizations). This focus on a top-down approach to SIW strategy and policy emphasizes strong government oversight over infrastructure sectors. These sectors constitute both the principal SIW vulnerability of the United States and the most prominent sources of SIW strength, because of private sector investments in defensive and encryption technologies. The two models of tactical warning and alert structures that are most plausible within this category are the NICON and the counterterrorism models. Each of these models offers a mix of mandates and incentives for the private sector in infrastructure protection, and each offers organizational guidance, with the federal government at the center of decisionmaking processes and information flows that are critical to protection of the United States from SIW assault.

### Club of SIW Elites

This end state posits that U.S. supremacy in SIW will be only historically transient. It also posits an environment in which only a small number of nations ("great

**Table B.1**

**Models of Tactical Warning and Alert Structure**

| | NICON Model | Counter/Terrorism Model | CDC Model | Industry-Led Model |
|---|---|---|---|---|
| Locus of rule making (public-private) | Public/private; Government-led, national security focus | Government-led, law-enforcement focus | Public-led; Private-sector support function | Private/public; Industry led |
| Principal actors | Sector emergency response centers (state and federal authorization)—private sector owners | Law enforcement agencies—intelligence community and military in supporting roles | Academics, industry, and government agencies | Commercial entities—government regulators |
| Promotion separated from oversight | No real separation—possible politicization of alert response | Pure oversight | Mixed | Weak public oversight—blurring of promotion and oversight roles |
| Professionalization of staff | High, especially in monitoring and analysis roles | Very high, with government certification | Very high, with government certification in collaboration with industry | Very high, with industry certification of standards |
| Mandates or advisories | Mandates | Industry mandates; advisories to the public | Advisories, with the potential for strong mandates | Mandates, and the establishment of compelling commercial standards |
| Structure (degree of hierarchy) | Confederated | Hierarchical authority—confederated response | Decentralized—strong coordination | Type 1: Hierarchical under monopoly conditions; Type 2: Fragmented under competition |
| Coupling of warning and alert response | Strong (MEII+) | Strong (MEII−) | Very strong (resilience focus) | Weak (emphasis on recovery and reconstitution) |
| Sanctions for noncompliance | Severe—criminal and civil liability | Severe—civil and criminal liability | Severe—civil and criminal liability | Insurance liability—some exposure to criminal liability |

powers") possess significant SIW capabilities.  In this setting,  the United States chooses to pursue a collaborative approach to SIW mitigation and consequence management.  Along with a relatively small group of like-minded countries, the United States would seek to reach agreement on nonuse of SIW tools and techniques. Although the logic of policy options supporting the continuation of this end state is similar to that described for the end state of U.S. supremacy, some major differences exist.  The role of the private sector in this model is more significant, even though it takes place under governmental guidance.

International collaboration on mitigating vulnerabilities and sharing information on attack patterns within the group require that information on defensive (and perhaps offensive) techniques be shared.  Because some of the most advanced knowledge on wide-area network protection resides in the private sector, this environment is more open than that of the U.S. supremacy environment to less-structured sharing of potentially sensitive SIW-related information among non-nation actors (private sector infrastructure operators and specialists in network protection).  The appropriate models for this end-state could be a mixture of those articulated above.  At the national level, the United States could deploy a warning and response system that has elements of both the counterterrorism and NICON models.  At the international or "alliance level," the United States, in concert with like-minded members of the SIW elite, might create an international police organization that has elements of the current anti–money laundering Financial Action Task Force (FATF) of the G-7.  Other international organizations might be created on the basis of the CDC model.

## Global Defense Dominance in SIW

The achievement of defense dominance in SIW would require extensive collaboration among all actors possessing significant offensive and defensive SIW capabilities. Again, because SIW vulnerabilities and strengths are both significantly a consequence of private sector decisions regarding anti-intrusion defenses and security, the public sector must share responsibility for tactical warning and alert structures with nongovernmental entities.  International collaboration in this area also is necessary because of the levels of interconnection prevalent within the GII.  Detection of intrusions and the broader characterizing of network vulnerabilities might be strong reasons for a less centralized—and hence less governmental—approach to tactical warning and alert functions.

Under this model, government might be satisfied with providing broad guidance to private sector (both corporate and academic institutions, for example) entities on the standards of safety, security, and information sharing that they would have to meet to guarantee a stated level of tactical warning and alert capability.  In this model as in the other models, the private sector has a significant role to play in early warning of intrusions—either for uncoordinated criminal activity or a more structured wide-area assault.  Because under this end state, national infrastructures are more or less invulnerable to attack caused by the extensive reconstitution focus of SIW protection, the SIW problem itself is less central to those charged with protecting national security.

Based on the technological and operational proposition that SIW could be rendered ineffective as a form of strategic warfare, the probable warning and response system would likely take on elements of the counterterrorism and CDC models. The strategy to sustain this end state over time would likely require a high degree of international cooperation and cohesion. Strong moral, political, and practical taboos to the use of SIW as an offensive weapon would have to be nurtured, if not strongly reinforced, by the threat of national and international sanctions. Maintaining the "technical condition" of defense dominance would require collaboration among those most technically attuned to mitigating SIW vulnerabilities. This end state model would also probably require an appropriate degree of international cooperation, if only to minimize duplication in tactical warning system coverage.

## Disruption Mitigation Through Market-Based Diversity

This end-state posits a radically diminished role for government in tactical warning and alert functions. A strategy pursuing this end state relies on the commercial marketplace to reduce the vulnerability of information-dependent infrastructures by virtue of the sheer diversity and inherent resilience of the global information infrastructure. In turn, SIW defensive investments would be developed and procured primarily through the unilateral defensive investments of the private sectors, as they sought to protect themselves from criminal threats such as theft, fraud and low level terrorism. An apt metaphor is the natural development of a biological immune system, in this case, the technological diversity and dynamism of the GII fosters are inherent robustness.

A critical assumption of this end state, which directly influences the selection of such an approach to tactical warning and alert response, is that individual investments in infrastructure-level protection aggregate into national SIW defenses of great robustness. Depending on one's confidence in this assumption, the government's role in tactical warning shifts from one of broad standard setting and information collection (on the state of events in the SIW environment) to a more directly involved mode in which government shapes the SIW defenses deployed by the private sector (under the rubric of infrastructure protection) through both mandates and incentives. National warning and response structures under this strategy would thus contain elements of both the CDC approach and the more purely industry-led models. At the GII level, the United States, in league with like-minded members of the global economic elite, might create an international police organization that has elements of the current anti–money laundering FATF of the G-7. Collaboration between national and multinational owners and operators of large information infrastructures might well create an international clearinghouse and regulatory organization that had features similar to those of the BIS.

It seems likely, however, that the degree of fidelity to this approach would be heavily influenced by historical experiences with infrastructure assaults through cyberspace. The more structured that infrastructure attacks appear, the more a decentralized approach to tactical warning and alert functions would be vulnerable to criticisms that it was an inadequate way of protecting U.S. SIW security assets.

## RESEARCH AND DEVELOPMENT INVESTMENT STRATEGIES AND PRIORITIES

As is the case with the tactical warning and alert structure issue, R&D decisions made in the private sector may have great significance for the aggregate robustness of information-dependent infrastructures against SIW attack. The government thus faces decisionmaking on the appropriate role that it can play in influencing the selection of technologies and sciences that the private sector chooses to explore in R&D on network vulnerabilities. R&D decisions involve many factors, depending on the actors involved (public or private sector, United States or international), the interests driving the choice of research area (basic scientific inquiry versus engineering or policy-driven applications), and the resource intensity of the enterprise (how much R&D costs, and what the distribution of costs is among collaborating entities). Each of these factors affects the influence that government can have over societal R&D priorities. Under the rubric of an overall strategy for influencing the SIW end state, however, government must make choices that balance these three factors in ways that serve the national security interest.

Based on the four models shown in Table B.2, four alternative courses of action in R&D decisionmaking are formulated. These are (1) national security–oriented network protection goals, (2) coordinated defensive R&D with allies, (3) international proscriptions on offensive SIW R&D, and (4) a private sector focus. National security priorities affecting SIW are fundamentally a subset of broader strategies aimed at countering perceived threats to the interests of the United States and its friends and allies. As a result, it is possible that decisions made in areas related to information-dependent infrastructures, for instance, decisions on international network standards and electronic commerce, could have significant impact on the ability of an SIW-driven R&D strategy to make sustained progress.[1]

### Option 1: National Security-Oriented Network Protection Goals

As with traditional defense-related R&D, government agencies would participate in a more or less formalized R&D program, making contracts available to private sector institutions (both commercial and academic) to augment the resources of the federal government. A sample rank-ordered list of R&D priorities under this model is

1. Information assurance

2. Monitoring and threat detection

3. Protection and mitigation

4. Vulnerability assessment and analysis

5. Risk management

6. Contingency planning

---

[1]For an in-depth discussion of options for R&D investments in the SIW domain, see Robert H. Anderson, and Anthony C. Hearn, *An Exploration of Cyberspace Security R&D Investment Strategies for DARPA*, Santa Monica, Calif.: RAND, MR-797-DARPA, 1996.

**Table B.2**

**Models of Infrastructure Protection Research and Development Resource Allocation Priorities**

| | National Security—Oriented Network Protection Goals | Coordinated Defensive R&D with Allies | International Proscriptions on Offensive SIW R&D | Private Sector Focus |
|---|---|---|---|---|
| | Option 1 | Option 2 | Option 3 | Option 4 |
| Proportion of government expenditures in aggregate infrastructure protection R&D | High, >50 percent | Moderate to high, approximately 50 percent, including collaborative international programs | Moderate, approximately 50 percent performed by the private sector | Low, < 50 percent of total spending deriving from the public sector |
| Decision-making process for R&D decisions | Political process  National security focus | International political consultations  Private sector R&D decisions | Public sector research on standards  Private sector business decisions | Private Sector  Market driven |
| International coordination of R&D resource allocation | Limited  Close allies only | Government-to government MoUs  International private sector joint ventures | Treaty-based definitions of permissible SIW-related R&D | Coordination through commercial joint ventures  Sharing of intellectual property among consortium partners |
| Principal R&D issue area priority | Information assurance | Information assurance  Monitoring and threat detection | Monitoring and threat detection | Risk management  Contingency planning |
| SIW R&D proscriptions | Legal restrictions on domestic private sector research on network security applications | Legal guidelines covering private sector INFOSEC R&D | Offensive SIW research banned | None |
| Balance of government expenditures for research (R) vs. development (D) | +R, +D | +R, −D | +R, negligible D | +R, zero D |

This list encompasses research designed to provide enhanced protection for government-specific information infrastructure assets, as well as those critical infrastructure resources owned and operated by the private sector. The priority on information assurance and on monitoring and threat detection illustrates a perceived necessity for a national information assurance program that places considerable emphasis on infrastructure protection policies in all critical information-dependent industrial and service sectors. Priorities in this area might include the coordination of information assurance programs between industry and governmental oversight agencies, the design of curricula and training programs for information assurance in industry, and the design of information systems and software for monitoring network traffic and intruder activity.

Items 3 through 6 in the above list are other activities important for ensuring the security and relative immunity of the United States to infrastructure (SIW) attack. For each of these areas, targeted government research projects would be designed to enhance United States capabilities across the board. In turn, other kinds of incentives could also be used to influence private sector R&D activities to support the broad immunity of U.S. infrastructures to strategic assault. Priorities within the SIW domain would, however, require a significant U.S. government commitment, with the knowledge that whatever resources were expended might still be dwarfed by those spent in the private sector for overlapping (and potentially conflicting) reasons.

## Option 2: Coordinated Defensive Research and Development with Allies

Coordinating research and development activities with key allies offers the opportunity for the United States to leverage its defensive SIW research against potentially useful skills and technologies held by the U.S. friends. This approach recognizes that awareness of the underlying defensive technologies of importance to SIW is unlikely to be held solely by the United States. Instead, major technological and economic competitors in Europe and Asia, for example, are likely to have significant capabilities of value to a defensive R&D program.

The of priorities for defensive SIW R&D would not change much within this category, but the activities of government may change. In particular, the private sector could provide services to multiple national actors on the basis of coordinated private sector R&D activities under the auspices of an international agreement. The commercial value of many of the underlying technologies might lend itself to an independent decision (by corporations) to conduct R&D in areas of importance to commercial products. This private sector's interest in product and technology development may require that government R&D strategy place greater emphasis on support for basic research in the network protection field.

Thus, the priorities for defensive SIW R&D resembles those of the defense-centered approach. However, these would be modified because governmental expenditures are more narrowly targeted toward basic research.

## Option 3:  International Proscriptions on Offensive SIW Technological Research and Development

This approach to R&D strategy and priorities may have the most traditional focus.  It depends on formal proscriptions on offensive SIW R&D, with collaborative enforcement of this prohibition provided through treaty-determined sanctions or collective decisionmaking.  The political environment thus sets broad constraints on the types of information network–related research that could be undertaken by governments. Defensive SIW R&D would still be allowed, though differentiating those activities from offensive research might be difficult.  Government priorities would be relatively passive detection and monitoring technologies, with active countermeasures limited by international agreement.  Thus, the resulting list of R&D priorities would be as follows:

1.  Monitoring and detection

2.  Protection and mitigation

3.  Information assurance

4.  Risk management

5.  Contingency planning

6.  Vulnerability assessment and threat detection

This approach also seems to favor a continuing government role in basic research on SIW topics, though this would be constrained by international restrictions on permissible areas of inquiry.  Commercial applications of network protection techniques would thus become the major focus of societal efforts, as the burgeoning market for telecommunications and computing technologies continued to drive the technical state of the art toward greater capabilities at the least cost.  Government research might be reduced to a "market awareness" function that focuses on the defensive applicability of emerging information infrastructure technologies, rather than on any particular technical solution for government procurement or standard setting.

The fundamental problem with this approach might be the great difficulty in separating defensive and offensive SIW technologies. Because the underlying technical principles are quite similar, separating them might require institutional adaptations, or governments' forswearing entirely the direct development of tools and techniques so that all nations would have access to the commercial market for networking and information assurance resources.

## Option 4:  Private Sector Focus

This approach is a purely industry-dominated R&D model.  In this setting, information infrastructure owner-operators would set broad technology priorities in response to the requirements of their business.  In turn, the providers of network security and infrastructure protection technologies would exist in a classic supply-demand relationship, with the marketplace setting the overall level of security-related R&D activity taking place at any given time.  In this model the government would play an advisory role, consulting with industry and infrastructure operators to ensure that public safety and security concerns were reflected in the overall R&D priorities set by the private sector. Private sector priorities for SIW defensive R&D research might include the following.

1. Risk management

2. Contingency planning

3. Public sector research on technical standards

4. Public sector coordination of monitoring and threat detection

A significant difference in this model is that R&D supports product and service delivery by the private sector, and only indirectly reflects public requirements for security and defensive immunity from structured attack. In essence, an assumption is made that an aggregate private sector R&D strategy aimed at security information–dependent infrastructures will cover all important areas of importance to defensive SIW protection. In the absence of a successful SIW attack by a foreign actor, this approach might be seen as persuasive, because it is a much less expensive (from the point of view of government budgeting) option than any of the previous models.

At the same time, if important gaps in SIW defensive research remain, this model may prove inadequate in addressing governmental concerns about national information infrastructure security, and the relative immunity of U.S. infrastructures from broad-based disruption through SIW attacks. Thus, an argument can be made for a residual, but targeted, role of the public sector in funding research in areas of lesser interest to commercial firms. Determining in advance what those areas are, however, may be as difficult as finding out whether commercially oriented defensive SIW R&D would be adequate to protect national security.

## ALTERNATIVE END STATE RESEARCH AND DEVELOPMENT INVESTMENT STRATEGIES

**U.S. Supremacy in SIW.** Emphasis on R&D option 1 (national security oriented network protection goals) with limited emphasis on option 2 (coordinated defensive R&D with allies).

R&D priorities for this end state reflect the concern with maintaining dominance in all areas of technology that are important to information networks. The government-directed focus of R&D decisionmaking reflects a traditional defense-centered model of weapon system research analogous to that which existed during the Cold War. Another similarity between this model and historical defense R&D patterns is that selected international partnering on technology development is also envisioned as a means of leveraging U.S. research against the unique scientific and technological capabilities of allied nations.

**Club of SIW Elites.** Emphasis on R&D option 2 (coordinated defensive R&D with allies) with parallel emphasis on R&D option 3 (international proscriptions on offensive SIW technological R&D).

R&D activities under this end state are a combination of defensively oriented technical analysis and international proscriptions or negotiations on restricting some kinds of research on offensive SIW tools and techniques. The close relationship between research efforts in the defensive and offensive areas creates the potential for disagreement about the definition of proscribed activities. In turn, an exploration of the technical vulnerabilities of systems would directly provide information on the offensive potential of actors possessing an understanding of how such weaknesses might be exploited. Coordinated R&D conducted with allies would allow for a more efficient division of labor in addressing known information infrastructure vulnerabilities. However, when the club of SIW elites contained potential adversaries (peer or near-peer) it is likely that restrictions on research collaboration, and potentially on information sharing, would impede internationally coordinated SIW activities.

**Global Defense Dominance in SIW.** Emphasis on R&D options 2 (coordinated defensive R&D with allies), 3 (international proscriptions on offensive SIW Technological R&D), and 4 (private sector focus).

Coordinated R&D on defensive options under this end state would focus on the mitigation of significant infrastructure vulnerabilities and on the robustness (and reconstitution) of critical systems during disruptive events. Constraining these research efforts would be parallel efforts at defining, and proscribing, research on offensive SIW tools and techniques. To preserve a defense-dominant SIW environment, or to move toward such an end state, it would be necessary to define a class of activities that constitute *precursors* of offensive work. Private sector activities in infrastructure protection constitute an essential part of this effort, because they would likely spend the most money on vulnerability mitigation. However, hedging against a possible "breakout" by a rogue actor in the SIW domain would require some residual knowledge of the offensive SIW strategic landscape. It is possible that research on defenses would supply the requisite knowledge for characterizing potential offensive threats posed by rogue actors. When this is not the case, however, difficult decisions would be necessary regarding sensitive research areas related to offensive capabilities.

**Disruption Mitigation Through Market-Based Diversity.** Emphasis on R&D options 3 (international proscriptions on offensive SIW technological R&D) and 4 (private sector focus).

Under this end state market-oriented decisionmaking on promising R&D areas in infrastructure protection would dominate the options followed to protect against SIW vulnerabilities. Because the principal information-dependent infrastructures are privately owned, sector operators have an interest in preventing criminal (or terrorist) targeting or misuse of their property. Therefore, infrastructures of importance to U.S. national security could be protected through the normal insurance and risk management activities of private business owners. Under such an approach, government concerns about the targeting of critical infrastructures would be met by informal consultations with industry on the nature of potential threats to infrastructure security, and by international negotiations on norms proscribing offensive SIW weapon research and use. These international proscriptions would have importance to the R&D activities carried out by private sector entities, because commercially driven research might have applications in offensive SIW operations. A reconciliation of a commercial infrastructure protection research agenda with international restrictions on permissible R&D would have to be conducted. This reconciliation process would take place through consultations between industry and government, but may result in legal restrictions on technical research that could be undertaken by the private sector.

# EXEMPLARY SIW SCENARIOS

This list of exemplary SIW tools and techniques provides examples of the increasingly sophisticated methods available for network intrusion. As open-standards–based networking spreads within society, the vulnerability of both networks and data may increase if appropriate protective measures are not taken.

---

### EXEMPLARY SIW WEAPONS[1]

Logic Bomb—Deliberately inserted program code causing a computer program or system to perform an operation departing from normal operating parameters. Such a weapon could destroy critical data or create a critical failure in an importing computing or communications system.

Trojan Horse—A program that appears to perform a legitimate and preauthorized function, but conceals a second (or more) covert function that performs unauthorized operations within a computing system or network.

Viruses—Code fragments (within a computer program) that reproduce by attaching themselves to another program. Such viruses can damage data, degrade system performance, or permit unauthorized system access.

Packet Sniffer—A program that monitors the data within an IP network searching for passwords or other proprietary data.

Packet Spoofer—A program that inserts falsely addressed packets within an IP network datastream to gain unauthorized access to a computer system and/or network.

GPS Fuzzer—A device that scrambles local reception of GPS timing signals, thereby inhibiting their use in navigation and other applications.

Flooding—A technique that uses automated calls to a particular circuit to deny access to authorized users.

---

[1]For an in-depth discussion of options for R&D investments in the SIW domain, see Anderson, Robert H., and Anthony C. Hearn, *An Exploration of Cyberspace Security R&D Investment Strategies for DARPA*, Santa Monica, Calif.: RAND, MR-797-DARPA, 1996.

# THE STRATEGIC NUCLEAR WARFARE FRAMEWORK PROBLEM

## OBJECTIVE

The objective of this appendix is to provide a means of comparing the problem of crafting an inaugural framework for the SIW-related strategy and policy issues (and preparing for an evolving series of such frameworks) with the comparable experience for the strategic threat posed by nuclear weapons. This appendix therefore presents what is believed to be a reasonable version of certain events and contexts in the history of Cold War strategic nuclear warfare (SNW)–related strategy and policy decisionmaking, up to the present time. This history includes both SNN-related strategy and policy issues presented (or not presented) at certain points in time, alternative courses of action on these issues that seem to have been considered at the time, and the actual decisions made, to the extent that they are clear (or ripe for supposition or reasoned speculation).

## A HISTORY OF KEY DIMENSIONS AND ASYMPTOTIC END STATES FOR STRATEGIC NUCLEAR WARFARE

There were six hypothetical end states in the revolutionary development of the nuclear weapon in 1945. These included

A. U.S. monopoly

B. U.S. offensive and defense superiority

C. Mutual assured destruction or mutual assured retaliation

D. Proliferated and proliferating nuclear capable countries

E. Virtual abolition or nuclear deemphasis

F. Abolition

See Table D.1. for an outline of the major features of each end state. It will be demonstrated that major events, technological and geostrategic, profoundly affected the Cold War framework of all possible End States. In essence, major events of that time profoundly affected the feasibility and desirability of future end states.

Table D.1

**Alternative Strategic Nuclear Warfare Asymptotic End States**

| Key Dimensions | A U.S. Monopoly | B U.S. with Overwhelming Offensive/ Defensive Superiority | C Mutual Assured Destruction | D Proliferated and Proliferating | E Virtual Abolition | F Abolition |
|---|---|---|---|---|---|---|
| Number of nuclear powers | 1 | 2 (U.S.-led bloc vs. Soviet-led bloc) | U.S.-led bloc vs. Soviet bloc or, comparable standoffs | E.g., 15–20+ by 2020 and rising | Perm5 + ? with small nos. | 0 |
| SNW defensive capability | — | High | High | Moderate | Moderate | Low |
| Size of at-the-ready arsenals | Small | Large | Large | Moderate | Small | 0 |
| Virtual nuclear arsenals | — | Weak | N/A | Moderate? | Large? | Moderate? |
| Inspection & enforcement (international cooperation) | — | Weak | Weak to moderate | Weak to moderate | Very extensive | Very extensive |
| Non-nuclear forces | — | — | High (flexible responsed [RMA]) | Moderate (RMA) | High (RMA+) | High (RMA) |
| Uncertainty in perpetrator identity | — | Low | Low | High | Low? | Low? |
| Tactical warning/attack assessment | — | Very high | Very high | Very high | ? | ? |

## Inaugural Period 1945–1948

At the beginning of the nuclear era, four end states were conceptualized (see Table D.2. The first was end state A, U.S. Monopoly, which was enshrined in the 1947 Atomic Energy Act. This act gave a civilian organization, the Atomic Energy Commission (AEC), peacetime control of the nascent nuclear arsenal. Furthermore, there were high hopes that a U.S. monopoly could be sustained for a decade or more through the rigorous denial of the science and technology gained during the Manhattan Project. The second was end state B, clear U.S. offensive and defensive superiority. At the time, the Joint Chiefs of Staff (JCS) viewed the nuclear arsenal as a new and more efficient means of conducting a strategic campaign against the Soviet Union in the event of World War III. Given the status of nuclear scarcity, that is, the small number of operational nuclear weapons, the JCS continued to believe that nuclear weapons would play an important but not decisive role in the next world war, which by 1947 was viewed as possible with the Soviet Union. Third was end

**Table D.2**

**Alternative Strategic Nuclear Warfare Asymptotic End States: Inaugural Period**
**(1945–1948)**

| Key Dimensions | A | B* | C | D | E |
|---|---|---|---|---|---|
| | U.S. Monopoly | | Mutually Assured Destruction | Proliferated and Proliferating | Abolition |
| Number of nuclear powers | 1 | | "2 party" (U.S.-led bloc vs. Soviet-led bloc) | e.g., eventually dozens | 0 |
| SNW defensive capability | — | | Low | ? | — |
| Size of at-the-ready arsenals | Small | | Moderate to large | Small to large | 0 |
| Inspection & enforcement (international cooperation) | — | | Moderate | Weak to moderate | Very extensive |

*Revolution in Military Affairs.

state C, mutual assured destruction or mutual assured retaliation. In 1946 and 1947, several civilian strategic thinkers, such as Bernard Brodie, made the case that nuclear proliferation was likely and that the civilian-destroying potential of nuclear warfare ensured that the weapons could be used only for deterrence. At that time, this concept was viewed as radical, although a version of this concept was raised among those informed who feared that nuclear weapon technology would rapidly proliferate and that the international environment would end up with end state D, a highly nuclear-proliferated world.

Motivated by fears that the an "atomic monopoly" was not sustainable, a major diplomatic initiative, the Acheson-Lillenthal plan, was pushed by the Truman administration to place all nuclear weapons and their associated infrastructure under an international authority, that is, end state E, abolition. By the end of this phase in 1948, the end state of abolition had all but disappeared, with the intensification of the political and geostrategic rivalry between the United States and the Soviet Empire, called the Cold War. End state F, virtual abolition, reflects the view that pure abolition would be imprudent, that some "lid on the nuclear jar," will be necessary to protect against breakout. That objective is presumably achievable through a small number of countries having very small nuclear arsenals (e.g., a few hundred nuclear weapons or even less).

## 1949–1954

Major events between 1949 and 1951 radically foreclosed two possible end states: A, a U.S. monopoly, and F, abolition. See Table D.3. The monopoly was shattered by the Soviet atomic test in 1949, the militarization of the Cold War prompted by the Korean Conflict, and the appearance of a Sino-Soviet alliance. In addition, the first event of nuclear proliferation beyond the Soviet Union occurred with the successful

Table D.3

**Alternative Strategic Nuclear Warfare Asymptotic End States:  1949–1954**

| Key Dimensions | B<br>U.S. with Overwhelming Offensive/Defensive Superiority | C<br>Mutually Assured Destruction | D<br>Proliferated and Proliferating |
|---|---|---|---|
| Number of nuclear powers | "2 party"<br>(U.S.-led bloc vs. Soviet-led bloc) | 2 party<br>(U.S.-led bloc vs. Soviet-led bloc) | E.g., eventually dozens |
| SNW defensive capability | High | High | Moderate |
| Size of at-the-ready arsenals | Large | Large | Moderate |
| Inspection & enforcement (international cooperation) | Weak | Weak to moderate | Weak to moderate |
| Non-nuclear forces | Moderate<br>(New Look) | High<br>(flexible response) | Moderate to high |
| Tactical warning/attack assessment | Very high | Very high | Very high |

British nuclear weapon test in 1951.  Within the United States there was broad commitment to sustaining end state B, U.S. offensive and defensive superiority. Prompted by the explosive increase in defense spending, the era of "atomic scarcity" was rapidly replaced by the era of nuclear weapon "plenty."  Furthermore, thermonuclear weapons were rapidly being developed, thereby radically increasing the city-destroying potential of both forces.

Some supported the idea that some form of mutually assured destruction or deterrence posture, end state C, was plausible.  This idea prompted the first moves by the two "superpowers" to consider some arms restraint strategies after the death of Stalin.  Even with the advent of thermonuclear weapons and early moves toward "détente," the focus of U.S. defense planning was to exploit the U.S. nuclear arsenal as an "asymmetric" response to the clear non-nuclear superiority of the Red Army in the European theater.  This focus included a major investment in strategic air defenses because the main transoceanic means of delivery was the newly deployed long-range jet bomber.  The "New Look" strategy consciously relied on the efficiency of nuclear weapons to limit budgetary consequences of a peacetime mobilization of military power by the United States.  By the end of this period, the Soviet leadership was following a similar nuclear emphasis strategy.

## 1962–1964

Doubts about the U.S. reliance on a nuclear emphasis strategy became increasingly clear with the rapid development of the ICBM and the rapid nuclear weapon prolif-

**Table D.4**

**Alternative Strategic Nuclear Warfare Asymptotic End States:  1962–1964**

| Key Dimensions | C<br>Mutual Assured Destruction | D<br>Proliferated and Proliferating | F<br>Virtual Abolition |
|---|---|---|---|
| Number of nuclear powers | "2 party" (U.S.-led bloc vs. Soviet-led bloc) | E.g., eventually dozens | Perm5 + ? with small #s |
| SNW defensive capability | High | Moderate | Moderate |
| Size of at-the-ready arsenals | Large | Moderate to high | Small |
| Virtual nuclear arsenals | N/A | N/A | Moderate to high |
| Inspection & enforcement (international cooperation) | Weak to moderate | Weak to moderate | Weak to Moderate |
| Uncertainty in perpetrator identity | Low | Moderate to high | Low? |
| Non-nuclear forces | High (flexible response) | Moderate | Moderate |
| Tactical warning/attack assessment | Very high | Very high | Very high |

eration by the Soviet Union.  See Table D. 4.  Offensive and defensive nuclear war-fighting strategies, end state B, were called into question during the psychologically traumatic Berlin and Cuban crises.  After the Cuban Missile Crisis in 1962, the geostrategic concept of détente with the Soviet Union developed, along with the notion that the nuclear arms competition could be stabilized by mutual agreement. This mutual arms control process would stabilize end state C, mutually assured destruction between the two superpowers.  Simultaneously, a major effort was made to slow down, if not stop, further nuclear weapon proliferation, end state D, after the successful French and Chinese nuclear weapon tests.  Both the French and Chinese successfully resisted the efforts of both superpowers to contain their programs. During this period, some strategic thinkers argued that the preferred long-term objective of the détente and arms control process was a radical reduction of both superpowers' nuclear arsenals to a few hundred weapons.  What then was known as "minimum deterrence" was advocacy for end state E, virtual abolition.  Unlike the mid-1950s, and late 1950s, support for a massive investment in strategic defenses or the maintenance of an offensive disarming capability, end state B, waned in the face of the Soviet nuclear and missile buildup.

## 1992–1997

With the abrupt end of the Cold War, the framework of possible nuclear weapon end states has dramatically expanded from the previously very constricted eras. See Table D. 5. A de facto variant of end state B which includes the asymmetric deployment of strategic nuclear defenses in the face of a gravely weakened  Russian Republic.

On the other hand, end state C, mutually assured destruction or mutually assured retaliation remains the dominant policy of the current Administration in dealing with the Russian Republic. Driven by the fear that a post Cold War security environment might lead to an acceleration toward end state D, a proliferated world, great emphasis has been given to containing proliferation through a wide range of arms control initiatives. As a hedge against further nuclear weapon proliferation, the Administration has proposed to modify the current nuclear posture and arms control posture to accommodate the deployment of robust theater aerospace defense, i.e., the 1997 U.S.-Russia Helsinki Agreement. Most dramatically has been the strong reappearance of support for end states E, virtual abolition, and F, abolition. With the collapse of the Warsaw Pact followed by the end of the Soviet Empire, the underlying and enduring rationale for nuclear weapons within much of the U.S. national security establishment has been undermined.

Now, there is wide spread discussion about the geostrategic, technical, and operational challenges of executing major initiatives which deemphasize, if not eliminate, all operational nuclear arsenals. Noteworthy is that the issue of nuclear weapon mobilization options, that is, the virtual nuclear arsenals resident within the worldwide diffusion of nuclear power and science infrastructures has taken on greater prominence.

**Table D.5**

**Alternative Strategic Nuclear Warfare Asymptotic End States:  Mature Outline 1997**

| Key Dimensions | B U.S. with Overwhelming Defense Superiority | C Mutually Assured Destruction | D Proliferated and Proliferating | E Abolition | F Virtual Abolition |
|---|---|---|---|---|---|
| Number of Nuclear Powers | Perm5 + ? | U.S.-led bloc vs. Soviet bloc or, e.g., within Perm5 + ?) | E.g., 15-20+ by 2020 and rising | 0 | Perm5 + ? with small #s |
| SNW defensive capability | High | High | Moderate | Low | Moderate |
| Size of at-the-ready arsenals | Large | Large | Moderate | 0 | Small |
| Virtual Nuclear Arsenals | N/A | N/A | Moderate? | Moderate? | Large? |
| Inspection & enforcement (international cooperation) | Weak | Weak to moderate | Weak to moderate | Very extensive | Very extensive |
| Non-nuclear forces | Moderate (RMA) | High (Flexible Response/RMA) | Moderate (RMA) | High (RMA) | High (RMA+) |
| Uncertainty in Perpetrator Identity | Very Low | Low | High | Low? | Low? |
| Tactical warning/attack assessment | Very high | Very high | Very high | ? | ? |